INFECTIOUS DISEASES AND MICROBIOLOGY

HUMAN CORONAVIRUSES, FROM OC43 TO SARS-COV-2

INFECTIOUS DISEASES AND MICROBIOLOGY

Additional books and e-books in this series can be found on Nova's website under the Series tab.

INFECTIOUS DISEASES AND MICROBIOLOGY

HUMAN CORONAVIRUSES, FROM OC43 TO SARS-COV-2

MOHAMAD HESAM SHAHRAJABIAN
WENLI SUN
AND
QI CHENG

nova
Medicine & Health
New York

Library of Congress Cataloging-in-Publication Data

Names: Hesam Shahrajabian, Mohamad, author. | Sun, Wenli, author. | Cheng, Qi (Full Professor, Biotechnology Research Institute, Chinese Academy of Agricultural Sciences), author. Title: Human coronaviruses : from OC43 to SARS-CoV2 / Mohamad Hesam Shahrajabian, Wenli Sun, Qi Cheng.
Description: New York : Nova Science Publishers, 2020. | Series: Infectious diseases and microbiology | Includes bibliographical references and index. | Summary: "The coronaviruses are ssRNA viruses that infect a wide range of mammalian and avian species; they are important causes of respiratory and enteric disease, encephalomyelitis, hepatitis, serositis and vasculitis domestic animals. In humans coronaviruses are one of several groups of viruses that cause the common cold. The genus Coronavirus together with the genus Torovirus from the family Coronaviridae; members of these two genera are similar morphologically. The Coronaviridae, Arteriviridae, and Roniviridae are within the order Nidovirales. Seven coronaviruses are known to infect humans, three of them are serious, namely, SARS (severe acute respiratory syndrome, China, 2002), MERS (Middle East respiratory syndrome, Saudi Arabia, 2012), and SARS-CoV-2 (2019-2020). SARS is caused by a coronavirus (SARS-CoV) which exists in bats and palm civets in Southern China. Its family is Coronaviridae, and its genus is Coronavirus. The most important groups who are at risk are family members in close contact with cases, health-care workers in close contact with cases, elderly and immune compromised individuals appear at increased risk. MERS-CoV is a zoonotic virus which can lead to secondary human infections. It is the sixth coronavirus that influences human. MERS-CoV is most likely derived from an ancestral reservoir bats. MERS outbreak was found in the Republic of Korea since 2015. Coronavirus entry is initiated by the binding of the spike protein (S) to cell receptors, specifically, dipeptidyl peptidase 4 (DDP4) and angiotensin converting enzyme 2 (ACE2) for MERS-CoV and SARS-CoV, respectively. The genome sequence analysis has shown that SARS-CoV-2 belongs to betacoronavirus genus, which includes Bat SARS-like coronavirus, SARS-CoV and MERS-CoV. On the basis of nucleic acid sequence similarity, the newly identified 2019-nCoV is a betacoronavirus. The RBD portion of the SARS-CoV-2 pike proteins has evolved to effectively target a molecular feature on the outside of human cells called ACE2, a receptor involved in regulating blood pressure. The SARS-CoV-2 spike protein was found so effective at binding the human cells. In SARS-CoV-2, M protein is responsible for the transmembrane transport of nutrient, the bud release and the formation of envelope, S protein, attaching to hose receptor ACE2, including two subunits S1 and S2. These diseases can be considered important models for emerging infectious diseases as it emerged from natural animal reservoirs. Early recognition, prompt isolation and appropriate supportive therapy are the main parameters in combating with these deadly infections"-- Provided by publisher.
Identifiers: LCCN 2020029135 (print) | LCCN 2020029136 (ebook) | ISBN 9781536182590 (paperback) | ISBN 9781536183184 (adobe pdf)
Subjects: LCSH: Coronaviruses. Classification: LCC QR399 .H47 2020 (print) | LCC QR399 (ebook) | DDC 614.5/92414--dc23
LC record available at https://lccn.loc.gov/2020029135
LC ebook record available at https://lccn.loc.gov/2020029136

Published by Nova Science Publishers, Inc. † New York

CONTENTS

PREFACE

The coronaviruses are ssRNA viruses that infect a wide range of mammalian and avian species; they are important causes of respiratory and enteric disease, encephalomyelitis, hepatitis, serositis and vasculitis domestic animals. Virus is a tiny agent, around one-hundredth the size of a bacterium which can infect cells of plants and animals. A coronavirus has spikes around its spherical body (corona=crown). Coronaviruses are enveloped, positive-sense, single-stranded RNA viruses of the family Coronaviridae. Corona virus was first identified in the 1960s and is recognized causes of mild respiratory tract infections in humans. The first two HCoVs, HCoV-229E and HCoV-OC43 have been known since the 1960s. The viruses are subdivided into four genera on the basis of genotypic and serological characters which are Alpha-, Beta-, Gamma, and Deltacoronavirus, and among them the first two genera are those which infect humans. Seven coronaviruses are known to infect humans, three of them are serious, namely, SARS (severe acute respiratory syndrome, China, 2002), MERS (Middle East respiratory syndrome, Saudi Arabia, 2012), and SARS-CoV-2 (2019-2020). SARS-CoV, and MERS-CoV belong to betacoronaviruses (betaCoVs). For enveloped viruses, a critical player in the entry process is the viral fusion protein as it mediates the membrane fusion reaction. The viral entry into target cells for coronaviruses is performed by the spike (S) envelope glycoprotein, which mediates both host cell receptor binding and membrane fusion.

Chapter 1

CORONAVIRUSES

The coronaviruses are ssRNA viruses that infect a wide range of mammalian and avian species; they are important causes of respiratory and enteric disease, encephalomyelitis, hepatitis, serositis and vasculitis domestic animals. In humans coronaviruses are one of several groups of viruses that cause the common cold. The genus *Coronavirus* together with the genus *Torovirus* from the family *Coronaviridae*; members of these two genera are similar morphologically. The *Coronaviridae*, *Arteriviridae*, and *Roniviridae* are within the order *Nidovirales*. Coronaviruses are roughly spherical enveloped particles, 120-160 nm in diameter, with characteristic fringe of 15-20 nm petal-shaped surface projections (peplomers). In a subset of betacoronaviruses a second, inner fringe of 5-7 nm surface projections is also seen. Coronavirus (CoV) particles as studied by cryo-electron tomography are homogeneous in size and distinctively spherical (enveloped outer diameter 86 ± 5 nm). The envelope exhibits an unusual thickness (7.8 ± 0.7 nm), almost twice that of a typical biological membrane. The nucleocapsid is helical and tightly folded to form a compact structure that tends to closely follow the envelope. Coronaviruses have the largest known RNA genomes which comprise 28-32 kb of positive sense, single-stranded RNA. All coronaviruses have four structural proteins in common, a large surface flycoprotein, a small envelope protein acids, integral membrane glycoprotein, and a phosphorylated nucleocapsid protein. Major criteria for classifying human virus families is shown in Table 1. The most important properties of Coronaviruses are presented in Table 2. Classification of *Nidovirales* accepted by the International Committee on Taxonomy of Viruses are shown in Table 3. Schematic structure of particles of members of the order *Nidovirales* is presented in Figure 1.

Table 1. Major criteria for classifying human virus families

Criterion	Basic of Classification
Type of genomic nucleic acid	DNA or RNA
Nucleic acid strandedness	ds, ss, partially ds
Sense of ss nucleic acid	+, -, - with ambisense
Capsid morphology	Icosahedral, helical, or complex
Envelope	Present or absent
Genome segmentation	Number of segments
Genomic structure	For example, type of RNA cap, location of structural genes or repeat sequences
Electron micrographic (EM) appearance	For example, bullet-shaped rhabdoviruses or star-shaped astroviruses
Size of virion and/or genome	For example, large-genome DNA viruses (e.g., poxviruses, herpesviruses) versus small-genome viruses (e.g., picornaviruses, parvoviruses, hepadnaviruses)
Nature of gene expression, including nature and number of mRNA transcripts	For example, use of genomic polyproteins (e.g., picornaviruses, flaviviruses); use of reverse transcriptase (e.g., retroviruses, hepadnaviruses); use of multiple $3^{/}$ nested genes (e.g., coronaviruses); use of RNA ambisense coding (e.g., arenaviruses, bunyaviruses)

Table 2. The most important properties of Coronaviruses

Virion is pleomorphic spherical 80 to 220 nm (coronaviruses); or disk, kidney, or rod shaped 120 to 140 nm (toroviruses).
Envelope with large, widely spaced club-shaped peplomer.
Tubular nucleocapsid with helical symmetry.
Linear, plus sense ssRNA genome 27 to 33kb, capped, polyadenylated, infectious; untranslated sequences at each end.
Three or four structural proteins: nucleoprotein (N), peplomer glycoprotein (S), transmembrane glycoprotein (M), sometimes hemagglutinin-esterase (HE).
Genome encodes 3 to 10 further non-structural proteins, including the RNA-dependent RNA polymerase made up of subunits cleaved from two polyproteins translated from the 5/-end.
Replicates in cytoplasm; genome is transcribed to full-length negative sense RNA from which is transcribed a 3/-coterminal nested set of mRNAs, only the unique 5/ sequences of which are translated.
Virions are assembled and bud into the endoplasmic reticulum and Golgi cisternae; release is by exocytosis.
Variant viruses arise readily, by mutation and recombination, and the use of different receptors influences the host range exhibited.

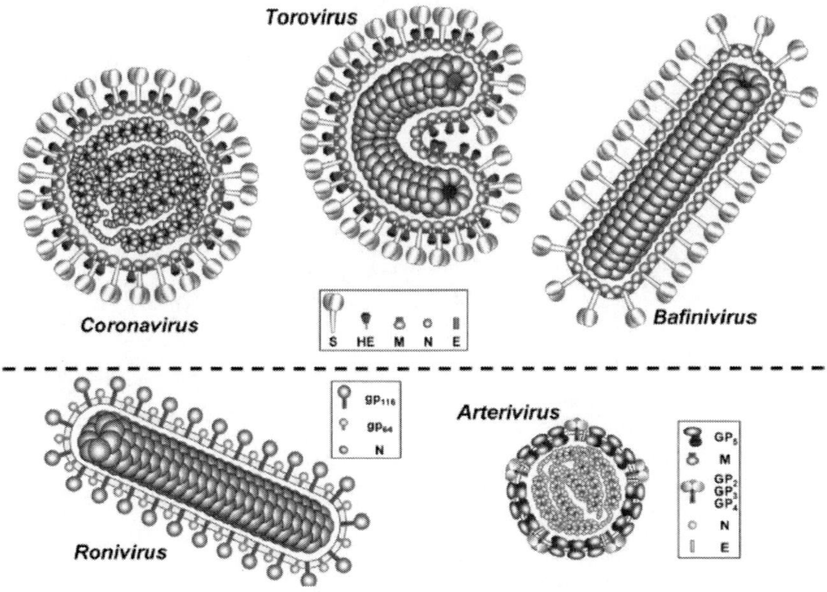

Figure 1. Schematic structure of particles of members of the order *Nidovirales*.

Table 3. Classification of *Nidovirales* accepted by the International Committee on Taxonomy of Viruses (ICTV)

Family	Subfamily	Genus	Type Species
Arteriviridae		Arterivirus	Equine arteritis virus (EAV)
Coronaviridae	Coronavirinae	Alphacoronavirus	Alphacoronavirus 1
		Betacoronavirus	Murine coronavirus
		Gammacoronavirus	Avian coronavirus
		Deltacoronavirus	Bulbul coronavirus HKU11
	Torovirinae	Torovirus	Equine torovirus
		Bafinivirus	White bream virus
Mesoniviridae		Alphamesonivirus	Alphamesonivirus 1
Roniviridae		Okavirus	Gill-associated virus

Chapter 2

HUMAN CORONAVIRUS OC43 (HCoV-OC43)

The first known coronavirus, the avian infectious bronchitis virus, was isolated in 1937 and was the cause of devastating infections in chicken. Coronaviruses are part of the coronaviridae family, which are positive-strand RNA viruses, and the coronaviridae subfamily is grouped into four genera, based on genetic differences and serological cross reactivity. There are findings showing that such putative enteric coronaviruses are antigenically unrelated to OC43 and 229E viruses (Luby et al., 1999). HCoVs were first identified in the 1960s with only two species known at the time, HCoV-229E and HCoV-OC43 (Jacomy and Talbot, 2003; Yang et al., 2010; Peters et al., 2015). Coronavirus and enterovirus/rhinovirus related acute respiratory illness were common. Older, chronically ill adults had more severe illnesses than young, healthy adults. Dyspnea was more common in older, chronically ill than in young, healthy adults. Respiratory illness symptom duration was longer in older, chronically ill adults. Older, chronically ill adults were more likely to receive antibiotics and steroids. Since 70% of all emerging infectious pathogens came from animals, the emergence of this novel virus may represent another instance of interspecies jumping of betacoronavirus from animals to human similar to the group A coronavirus OC43 possibly from a bovine source in the 1980s and group B SARS coronavirus in 2003 from bat to civet and human (Chan et al., 2012; Li and Lin, 2013; Killerby et al., 2018). Incident of infection is the highest in winter and early spring (Lim et al., 2016). HCoV-OC43 causes a common cold in humans BCoV causes gastrointestinal and respiratory tract disease in cattle (Hick et al., 2012; Bidokhti et al., 2013; Bok et al., 2015). BCoV exhibits a relative genetics stability when compared to HCoV-OC43, and the numerous recombination detected between HCoV-OC43 were much less frequent for BCoV (Kin et al., 2016). Human coronaviruses before COVID-19 is listed in Table 1. Classification of coronavirus is indicated in Table 2. Classification, discovery, cellular response and natural hose of the coronaviruses before COVID-19 is shown in Table 3.

Table 1. Human coronaviruses before COVID-19

Virus	Genus	Disease	Discovered
CoV-229E	Alpha	Mild respiratory tract infection	1967
CoV-NL-63	Alpha	Mild respiratory tract infection	1965
CoV-HKU-1	Beta	Mild respiratory tract infection; pneumonia	2005
CoV-OC43	Beta	Mild respiratory tract infection	2004
SARS-CoV	Beta	Human severe acute respiratory syndrome, 10% mortality rate	2003
MERS-CoV	Beta	Human severe acute respiratory syndrome, 37% mortality rate	2012
SARS-CoV-2	Beta	Severe acute respiratory infections, <2% mortality rate	2019

Table 2. Classification of coronavirus

Genera	Subgenera	Species
Alphacoronavirus	Colacovirus	Bat coronavirus CDPHE15
	Decacovirus	Bat coronavirus HKU10
		Rhinolophus ferrumequinum alphacoronavirus HuB-2013
	Duvinacovirus	Human coronavirus 229E
	Luchacovirus	Lucheng Rn rat coronavirus

Genera	Subgenera	Species
	Minacovirus	Ferret coronavirus
		Mink coronavirus 1
	Minunacovirus	Miniopterus bat coronavirus 1
		Miniopterus bat coronavirus HKU8
	Myotacovirus	Myotis ricketti alphacoronavirus Sax-2011
	Nyctacovirus	Nyctalus velutinus alphacoronavirus SC-2013
	Pedacovirus	Porcine epidemic diarrhea virus
		Scotophilus bat coronavirus 512
	Rhinacovirus	Rhinolophis bat coronavirus HKU2
	Setracovirus	Human coronavirus NL63
		NL63-related bat coronavirus strain BtKYNL63-9b
	Tegacovirus	Alphacoronavirus 1a
Betacoronavirus	Embecovirus (lineage A)	Betacoronavirus 1
		China Rattus coronavirus HKU24
		Human coronavirus HKU1
		Murine coronavirus
	Sarbecovirus (lineage B)	Severe acute respiratory syndrome-related coronavirus
	Merbecovirus (lineage C)	Hedgehog coronavirus 1
		Middle East respiratory syndrome-related coronavirus
		Pipistrellus bat coronavirus HKU5
		Tylonycteris bat coronavirus HKU4
	Nobecovirus (lineage D)	Rousettus bat coronavirus GCCDC1
		Rousettus bat coronavirus HKU9
	Hibecovirus	Bat Hp-betacoronavirus Zhejiang 2013
Gammacoronavirus	Cegacovirus	Beluga whale coronavirus SW1
	Igacovirus	Avian coronavirus
Deltacoronavius	Andecovirus	Wigeon coronavirus HKU20
	Buldecovirus	Bulbul coronavirus HKU11
		Coronavirus HKU15
		Munia coronavirus HKU13
		White-eye coronavirus HKU16
	Herdecovirus	Night heron coronavirus HKU19
	Moordecovirus	Common moorhen coronavirus HKU21

Table 3. Classification, discovery, cellular response and natural hose of the coronaviruses before COVID-19

hCoV genera	Coronaviruses	Discovery	Cellular receptor	Natural Host(s)
α-Coronaviruses	HCoV-229E	1966	Human aminopeptidase N (CD13)	Bats
	HCoV-NL63	2004	ACE2	Palm Civets, Bats
β-Coronaviruses	HCoV-OC43	1967	9-O-Acetylated sialic acid	Cattle
	HCoV-HKU1	2005	9-O-Acetylated sialic acid	Mice
	SARS-CoV	2003	ACE2	Palm Civets
	MERS-CoV	2012	DPP4	Bats, Camels

Coronavirus OC43 usually result in upper respiratory tract infections such as common cold; it may also cause severe pneumonia in patients presenting with comorbidities, but clinical signs alone do not allow for viral identification (Wu et al., 2003; Vandroux et al., 2018; Szczepanski et al., 2019). These viruses may be associated with the development of neurological diseases such as encephalitis (Morfopoulou et al., 2016). A hemagglutininesterase (HE) glycoprotein gene is only present in group 2 coronaviruses, which include human coronavirus OC43 (HCoV-OC43), bovine coronavirus (BCoV), porcine hemagglutinating encephalomyelitis virus (PHEV), canine respiratory coronavirus (CRCoV), mouse hepatitis virus (MHV), rat sialodacryoadenitis virus (SDAV) and equine coronavirus (ECoV) (Spaan et al., 1988; Zhang et al., 1992). It has reported that HCoV-OC43 and HCoV-229E are responsible for 10 to 30% of all common colds, and infections occur mainly during the winter and early spring (Larson et al., 1980). The evolutionary rate of the HCoV-OC43/BCoV pair was estimated in the order of 10^{-4} nucleotide substitutions per site per year (Vijgen et al., 2005), which is in the same range as reported for other RNA viruses (Domingo and Holland, 1988). The prototype HCoV-OC43 stain is, however, a laboratory strain, that since its isolation in 1967 was passed seven times in human embryonic tracheal organ culture, followed by 15 passages in suckling mouse brain, and an unknown number of passages in human rectal tumor HRT-18 cells and Vero cells; moreover, during the passage history, it is likely that a number of mutations have accumulated (Vijgen et al., 2005). St-Jean et al., (2004) reported that the complete genome sequence of a HCoV-OC43 clinical isolate, designated Paris. HCoV-OC43 Paris isolate might have been a result of cross-contamination with the ATCC HCoV-OC43 strain (Vijgen et al., 2005). Vandroux et al., (2018) reported that OC43 infection cannot be detected on the sole basis of clinical signs, and the presence of underlying comorbidities impacted the clinical outcomes of OC43 infections, similar to what is observed in the ongoing MERS-CoV infection in the Middle East. On the basis of phylogenetic analysis based on complete genome of all available coronaviruses consistently show that CRCoV BJ232 is most closely related to human coronavirus OC43 (HCoV-OC43), and BCoV, forming a separate clade that split off early from other Betacoronavirus 1 (Lu et al., 2017). Zhang et al. (1992) indicated that on the basis of phylogenetic analysis, hemagglutinin/esterase (HE) genes of coronaviruses and influenza C virus have a common ancestral origin, and that bovine coronaviruses and HCV-OC43 are closely related. Kunel and Herrler (1993) observed a high degree of sequence homology between the S proteins of HCV-OC43 and bovine coronavirus. Neighbor joining phylogenetic tree of the complete spike gene nucleotide sequence data of group 2 coronaviruses is shown in Figure 1. Consensus tree for cDNA sequences from a 251-nucleotide region of the polymerase gene of 12 coronaviruses is presented in Figure 2. Nucleotide alignment of qPCR primer binding sites. Nucleotide sequences of the (A) CRCoV NF3 and (B) CRCoV NR4 primer-binding sites is indicated in Figure 3.

Butler et al. (2006) found that some coronaviruses such as HCoV-OC43 and SARS-CoV adapt to growth in cells from heterologous species and this adaptability has facilitated the isolation of HCoV-OC43 viral variants with markedly differing abilities to infect animals and tissue culture cells. HCoV-OC43 showed increasing neurovirulence with passage through the murine brain McIntosh et al., 1967).

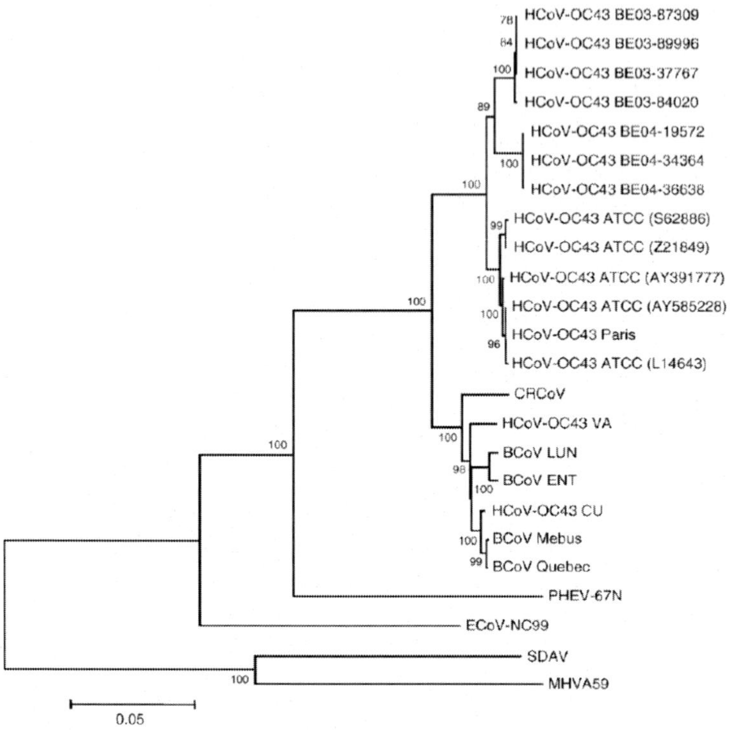

Figure 1. Neighbor joining phylogenetic tree of the complete spike gene nucleotide sequence data of group 2 coronaviruses: HCoV-OC43 BE03 strain (GenBank accession numbers AY903454, AY903456, AY903457, AY903459), HCoV-OC43 BE04 strain (AY903455, AY903458, AY903460), HCoV-OC43 ATCC VR759 strain (AY391777, AY585228, L14643, Z21849, S62886, Z32768, Z32769), HCoV-OC43 serotype OC43 Paris (AY585229), BCoV LUN (AF391542), BCoV ENT (AF391541), BCoV Mebus (U00735), BCoV Quebec (AF220295), CRCoV (AY150272), PHEV strain 67N (AY078417), ECoV stain NC99 (AY316300), SDAV (AF207551), and MHV strain A59 (AY700211). Bootstrap values over 75% are shown (Vijgen et al., 2005).

Unlike other coronaviruses, HCoV-OC43 and the closely related bovine coronavirus (BCoV) appear to bind to cells via *N*-acetyl-neuraminic acid (Schultze and Herrler, 1992; Vlasak et al., 1988). Risku et al., (2010) found that the significance of coronaviruses as gastrointestinal pathogens in children appears minor, since most of the coronavirus findings were co-infections with known gastroenteritis viruses.

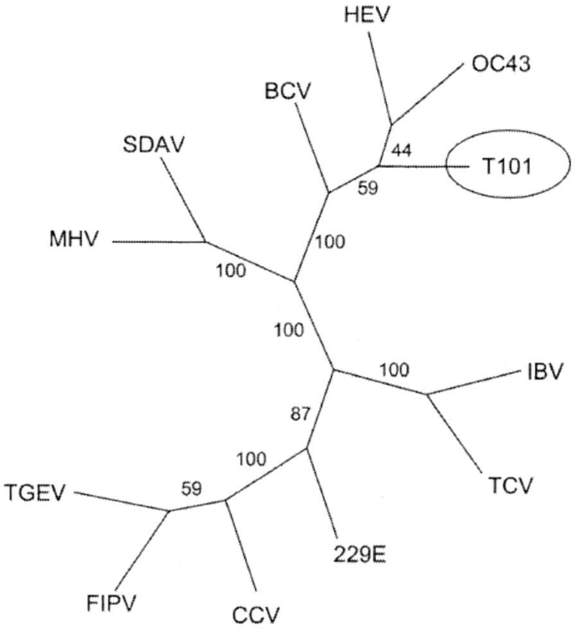

Figure 2. Consensus tree for cDNA sequences from a 251-nucleotide region of the polymerase gene of 12 coronaviruses. The sequence obtained from the canine respiratory coronavirus is designated T101. The numbers indicate bootstrap values obtained by analysis of 100 data sets. BCV, bovine coronavirus; CCV, canine coronavirus; FIPV, feline infectious peritonitis virus; HEV, hemagglutinating encephalomyelitis virus; IBV, infectious bronchitis virus; MHV, mouse hepatitis virus; OC43, human coronavirus strain OC43, SDAV, sialodacryoadenitis virus; TCV, turkey coronavirus; TGEV, transmissible gastroenteritis virus; 229E, human coronavirus strain 229E; T101, canine respiratory coronavirus (PCR product from tracheal sample T1010) (Erles et al., 2003).

Jacomy et al., (2006) suggested that viral persistence could be associated with an increased neuronal degeneration leading to neuropathology and motor deficits in susceptible individuals. Liang et al., (2013) reported three different patterns for the immunoreactivities of the three structural regions of HCoV-OC43 NP in human serum, which showed variability in the immune responses that occur during HCoV-OC43 infection in humans; the central0linker region of the NP appeared to be the most highly immunoreactive region for all three patterns observed. Four major protein species were resolved by electrophoresis and many of their properties were deduced from digestion studies using proteolytic enzymes, the four proteins are: (1) A 190 kda protein, the presumed peplomeric protein, that was glycosylated and proteolytically cleavable by trypsin into subunits of 100 and 90 kDa, (2) a 130 kDa protein that was glycosylated and behaved as a disulfide-linked dimer of 65 kDa molecules, (3) a 55 kDa nucleocapsid protein that was phosphorylated; (4) a 26 kDa matrix protein that was glycosylated; moreover, the 190, 130, 55 and 26 kDa species can therefore be designated P, H, N and M, respectively (Hogue and Brian, 1986). CryoEM structure of the apo-HCoV-OC43 S glycoprotein is shown in Figure 4. Structural studies of human CoV attachment to host receptors is indicated in Figure 5.

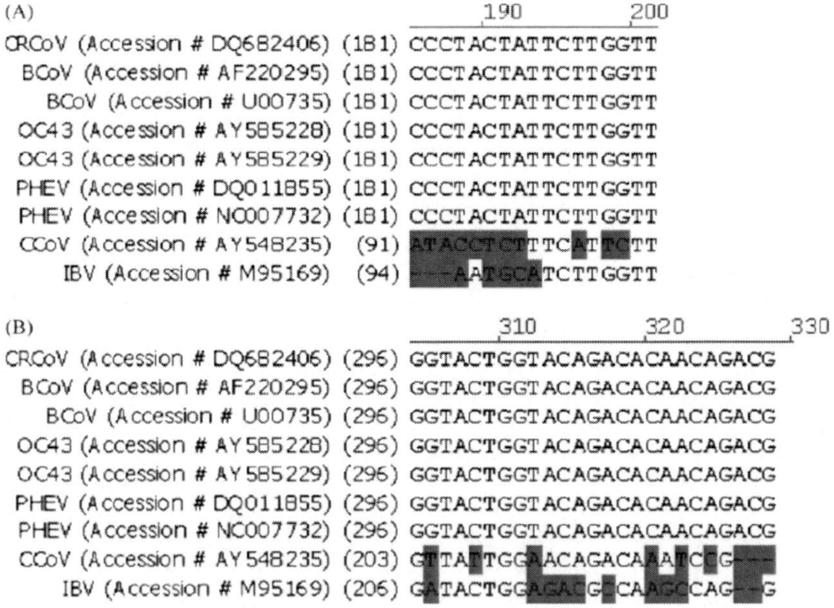

Figure 3. Nucleotide alignment of qPCR primer binding sites. Nucleotide sequences of the (A) CRCoV NF3 and (B) CRCoV NR4 primer-binding sites. The alignment includes sequences from the group 2 coronaviruses CRCoV, BCoV (bovine coronavirus), OC43 (human coronavirus OC43), PHEV (porcine hemagglutinating encephalomyelitis virus), group 1 coronavirus CCoV (canine coronavirus), and group 3 coronavirus IBV (infectious bronchitis virus). The numbers in the brackets show the starting position of the primer binding site within each sequence. The numbers along the top show the nucleotide positions within the alignment. Shaded areas indicate the location of the non-conserved nucleotides in CCoV and IBV (Mitchell et al., 2009).

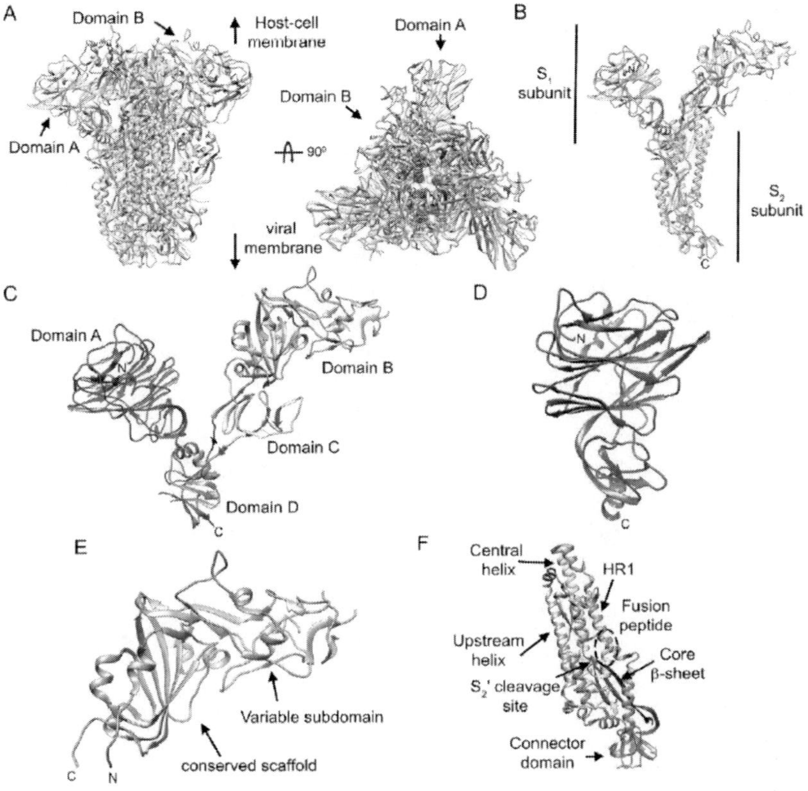

Figure 4. CryoEM structure of the apo-HCoV-OC43 S glycoprotein. (A) Ribbon diagrams of the apo HCoV-OC43 S ectodomain trimer (PDB: 6OHW) in two orthogonal orientations, from the side (left) and from the top, looking towards the viral membrane (right). (B) Side view of one S promoter. (C) Ribbon diagram of the HCoV-OC43 S_1 subunit. (D-E) Close-up view of HCoV-OC43 domain A (D) and domain B (E). (F) Ribbon diagram of the HCoV-OC43 S_2 subunit in the prefusion conformation. The N- and C-termini are labeled in panels (B-E).

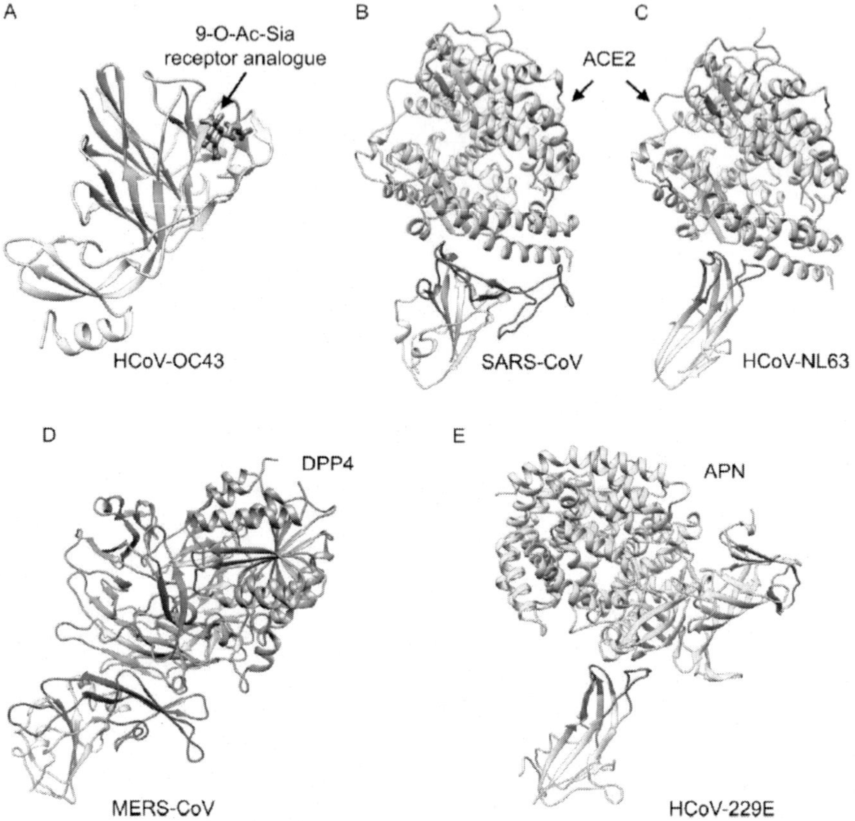

Figure 5. Structural studies of human CoV attachment to host receptors. (A-E), Ribbon diagrams of the complex between domain A of HCoV-OC43 S with a 9-*O*-Ac-Sia receptor analogue ((A) PDB: 6NZK), or the domain B of SARS-CoV S with ACE2 ((B) PDB: 2AJF), HCoV-NL63 S with ACE2 ((C) PDB: 3KBH), MERS-CoV S with DPP$ ((D) PDB: 4L72) and HCoV-229E S with APN ((E) PDB: 6ATK). In panels (B-E), each domain B is rendered in light blue and the receptor binding-motifs are colored purple.

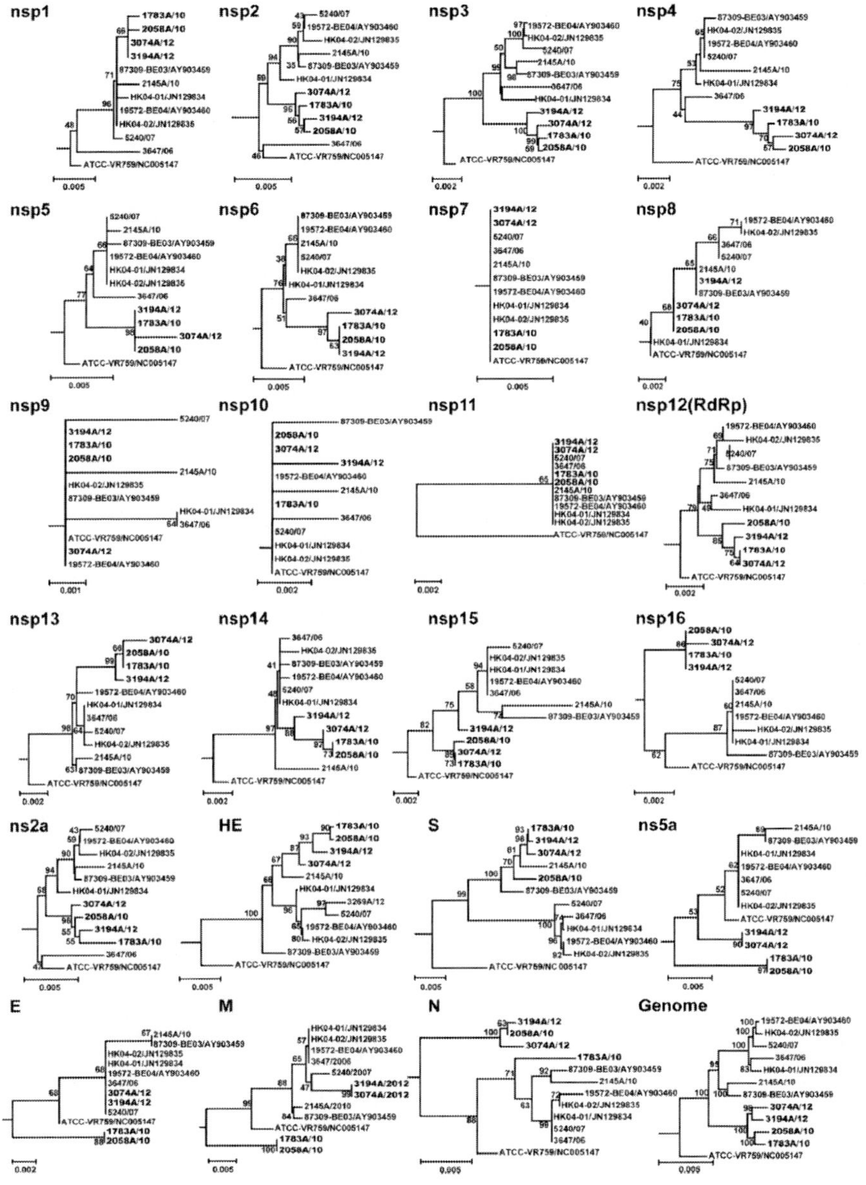

Figure 6. Phylogenetic analysis of multiple gene regions of HCoV-OC43 strains. A total of23 gene regions are analyzed, including nsp1 to nsp16, ns2α, E, M and N of ten genomes of HCoV-OC43. The neighbor-joining method (Kimura's two-parameter) was used to construct the trees with 1000 bootstrap values (Zhang et al., 2015).

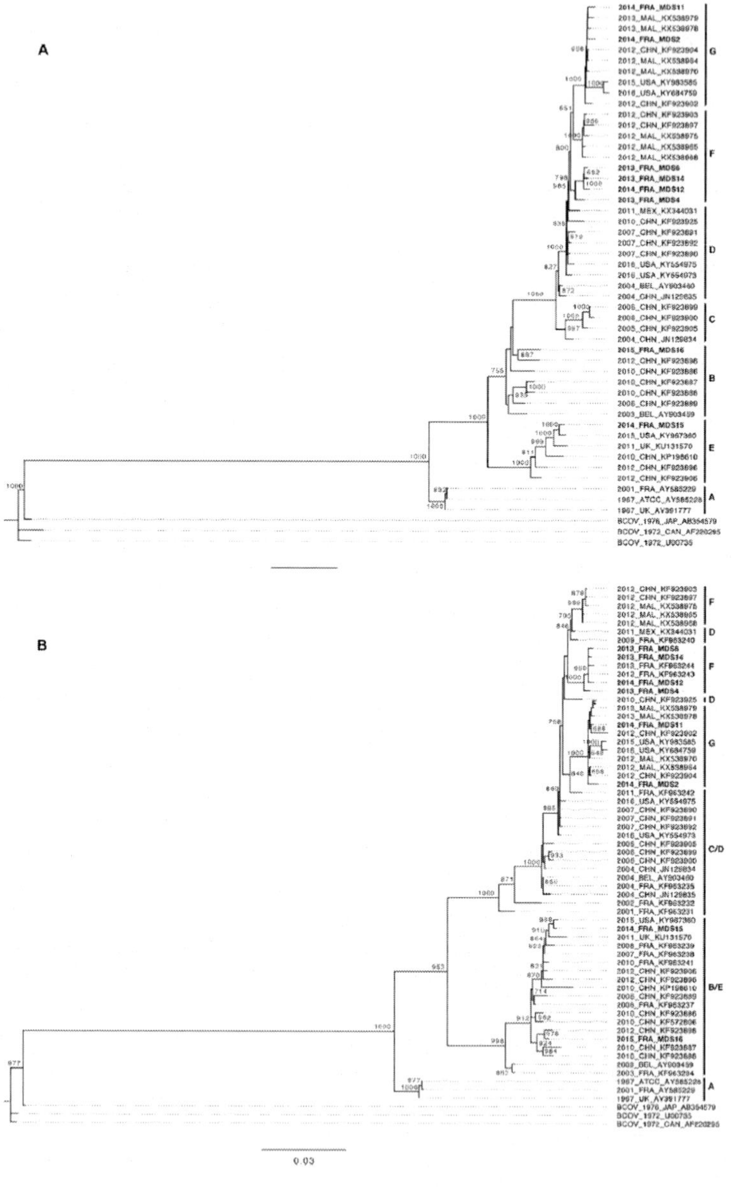

Figure 7. Maximum-likelihood trees of HCoV-OC43 strains. A, B, C, D, E, G genotypes/clades are represented. MDS sequences are in bold. Tree obtained from whole-genome sequences. B. Tree obtained from S gene sequences (Beury et al., 2020).

Cov-OC43 nucleoprotein antigens was introduced to the mariPOC test in 2017, the mariPOC respi test is able to detect nine respiratory viruses (influenza A and B viruses, respiratory syncytial virus, adenovirus, human metapneumovirus, para-influenzavirus 1-3, human bocavirus, and *Streptococcus pneumonia* from one nasopharyngeal sample at the point of care (Bruning et al., 2018). Phylogenetic analysis of multiple gene regions of HCoV-OC43 strains is shown in Figure 6. Schwarz et al., (2010) demonstrated that emodin can inhibit the 3a ion channel of coronavirus SARS-CoV and HCoV-OC43 as well as virus release from HCoV-OC43 with a $K_{1/2}$ value of about 20 μM; which suggests that viral ion channels may be a good target for the development of antiviral agents. HCV-OC43 and BCV might have diverged from each other fairly recently and that the 3′-end of the leader sequence has significant functional roles (Kamahora et al., 1989). Beury et al., (2020) identified genotypes B, E, F and G, two clusters of patients were defined from chronological data and phylogenetic trees. Analyses of amino acids substitutions of the S protein sequences identified substitutions specific for genotype F strains circulating among European people. Chen et al., (2013) showed that the strength of N protein/RNA interactions is critical for HCoV-OC43 replication. Patrick et al., (2006) confirmed that cross-reactivity to antibody against nucleocapsid proteins from CoV-OC43 must be considered when interpreting serological tests for SARS-CoV. Beidas and Chehadeh (2018) reported the transcriptional activity of ISRE, IFN-β promoter, and NF-κB-RE was significantly reduced in the presence of HCoV-OC43 ns2a, ns5a, M, or N protein, following the challenge of cells with Sendai virus, IFN-α or tumor necrosis factor-α, the expression of antiviral genes involved in the type I IFN and NF-κB signaling pathways was also downregulated in the presence of HCoV-OC43 structural or accessory protein. They have concluded both structural and accessory HCoV-OC43 proteins are able to inhibit antiviral response elements in HEK-293 cells, and to block the activation of different antiviral signaling pathways. Favreau et al., (2009) showed the importance of discrete molecular viral S determinants in virus-neuronal cell interactions that lead to increased production of viral proteins and infectious particles, enhanced UPR activation, and increased cytotoxicity and cell death. It has been found that the structure of HCoV-HKU1 provides a high-quality model for group 2A CoVs, which are distinct from group 2B CoVs such as severe acute respiratory syndrome CoV; the structure, together with activity assays, supports the relative conservation at the P1 position that was discovered by sequencing the HCoV-HKU1 genome (Zhao et al., 2008). Maximum-likelihood trees of HCoV-OC43 strains. A, B, C, D, E, G genotypes/clades are represented in Figure 7. The structure and topology of the HCoV-OC43 N-NTD is shown in Figure 8.

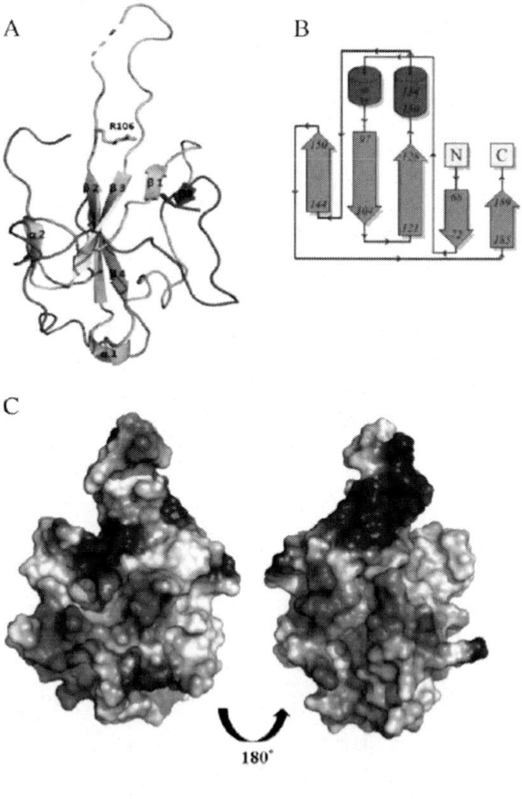

Figure 8. The structure and topology of the HCoV-OC43 N-NTD. (A) A ribbon diagram of the HCoV-OC43 N-NTD depicts the presence of five β strands, two 3_{10} helices, and several disordered regions. (B) The topology of the HCoV-OC43 N-NTD shows the relative positions of the secondary structures of the truncated protein. (C) The surface charge distribution of the HCoV-OC43 N-NTD (Chen et al., 2013).

Table 4. Differential nucleotide composition among coronaviruses

Coronavirus	ID	A	U	C	G
MERS	JX869059	26.2	32.5	20.3	20.9
SARS	NC_004718	28.5	30.7	20.0	20.8
229E	KF514433	27.1	34.7	16.6	21.6
OC43	NC_005147	27.6	35.6	15.2	21.7
NL63	JX504050	26.3	39.2	14.4	20.1
HKU-1C	DQ415912	27.8	40.1	13.0	19.1
HKU-1A	DQ415914	27.9	40.2	13.0	19.0
HKU-1B	DQ415911	27.7	40.3	12.9	19.1

Berkhout and van Hemert, 2015.

Table 5. Nucleotide and amino acid similarities of the major HCoV-OC43 (ATCC VR759) ORFs with the ORFs of BCoV, CRCoV, PHEV, ECoV, MHV and SDAV

HCoV-OC43 ORF				%Nucleotide (amino acid) similarity		
	BCoV	CRCoV	PHEV	ECoV	MHV	SDAV
ORF1a	97.4 (97.0)	NAa	NA	NA	69.3(65.9)	NA
ORF1b	97.8 (98.6)	NA	NA	NA	82.7(87.1)	NA
ns2	95.1 (95.0)	NA	NA	NA	59.0(59.7)	60.3(51.5)
HE	96.7 (95.3)	NA	89.8(88.7)	73.7(72.9)	60.8(58.1)b	64.2(59.5)
S	93.5 (91.4)	92.8(89.9)	81.7(81.0)	79.3(79.2)	66.5(62.9)	67.0(64.4)
ns12.9	96.1 (93.6)	NA	89.7(86.2)	88.8(81.7)	55.0(47.4)	59.8(51.4)
E	98.0 (96.4)	NA	99.6(98.8)	96.1(91.7)	72.5(65.9)	69.2(66.7)
M	94.8 (94.3)	NA	92.1(93.0)	91.2(88.7)	77.9(83.5)	77.5(82.3)
N	96.8 (96.4)	NA	94.3(94.0)	NA	71.8(69.6)	72.0(69.9)
Ia/Ibc	97.1 (NDd)	NA	95.2(ND)	NA	72.3(ND)	71.4(ND)

[a] NA, not applicable; the corresponding sequence is not available in the GenBank database.

[b] The 5′ end of the MHV HE ORF is missing due to a frameshift mutation or sequencing error in NC_001846.

[c] The OC43 (ATCC VR759 strain) internal ORF (I) coding region contains a stop codon at position 29345, resulting in two potential coding regions of 60 aa (Ia) and 115 aa (Ib). This stop codon is not present in BCoV, which has the capacity to code for a 207-aa protein. This stop codon is also absent in PHEV, MHV, and SADV.

[d] ND, not done.

Vijgen et al., 2005.

Differential nucleotide composition among coronaviruses is indicated in Table 4. Hypothetical model for the interaction of HCoV and BCoV virions with multivalent glycoconjugates. Schematically depicted are options of HCoV (OC43 or HKU1) and BCoV virions, with large spikes comprised of S (in gray), and smaller protrusions comprised of HE (yellow) extending from the viral membrane (blue) is presented in Figure 10. Nucleotide and amino acid similarities of the major HCoV-OC43 (ATCC VR759) ORFs with the ORFs of BCoV, CRCoV, PHEV, ECoV, MHV and SDAV is indicated in Table 6.

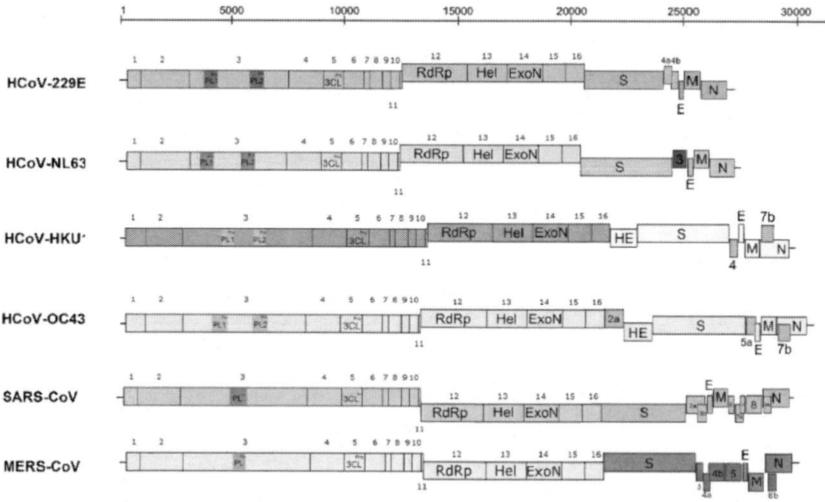

Figure 9. Schematic diagram of human coronavirus genomes. ORF1a/b, occupying the 5′ two thirds of the genome, translates into two replicase polyproteins pp1a and pp1ab as a result of ribosomal frameshifting, which are further cleaved to form 16 non-structural proteins nsp 1-16. This is followed by the subgenomic RNA fragments encoding the structural and accessory proteins. The genome structure of coronaviruses follow a characteristic order as shown in this diagram. ORF, open readin frame; PLpro, papin-like protease; 3CLpro, chymotrypsin-like protease; RdRp, RNA-dependent RNA polymerase; Hel, RNA helicase; RFS, ribosomal frameshift region; S, spike; E, envelope; M, membrane; N, nucleocapsid.

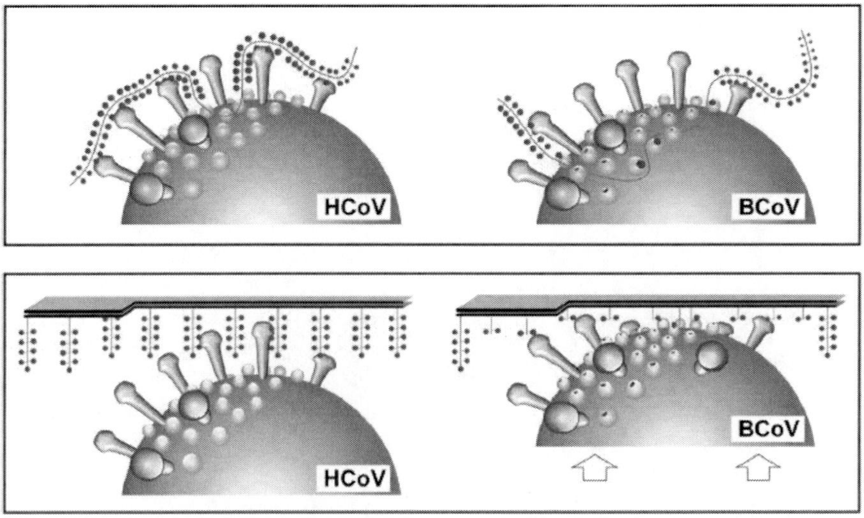

Figure 10. Hypothetical model for the interaction of HCoV and BCoV virions with multivalent glycoconjugates. Schematically depicted are options of HCoV (OC43 or HKU1) and BCoV virions, with large spikes comprised of S (in gray), and smaller protrusions comprised of HE (yellow) extending from the viral membrane (blue). Functional HE lectin CBSs are indicated by black holes (only one shown per HE dimer for reasons of simplicity). Also shown schematically are membrane-anchored (bottom) and non-anchored (top) mucin-type glycoconjugates of bottle-brush filamentous appearance with clustered receptors (9-O-Ac-Sias, red dots) arranged in linear arrays and with absence of red dots indicating receptor-destruction. The model based on the size difference between S and HE, visualizes how loss of HE lectin function might alter virion-assocaited receptor destroying activity, reducing the specific receptor destroying activity, reducing the specific activity of virions as well as the rate and selectivity of receptor destruction. In virions with lectin-deficient HEs, clustered substrates will be largely kept at a distance from the HE esterase catalytic pocket as a result of S-glycoconjugate interaction. In contrast, HEs with intact lectin CBS may draw in portions of the glycoconjugates (or will draw the virion-associated HEs toward them), aided by cooperativity of binding between adjacent HEs within the viral envelope. Thus, clustered glycotopes become fixed within reach of the esterase catalytic sites and receptor destruction is promoted (Bakkers et al., 2017).

Chapter 3

Human Coronavirus HKU1 (HCoV-HKU1)

HCoV-HKU1 was first isolated in 2005 from a 71-year-old man with pneumonia, and then several successive reports confirmed retrospectively that HCoV-HKU1 was circulating worldwide (Cui et al., 2010; Jin et al., 2010). The reported incidences varied from 0 to 4.4% of patients hospitalized for acute pulmonary and extrapulmonary symptoms, and laboratory detection is mostly achieved by RT-PCR (Chang et al., 1999; Pierangeli et al., 2007; Chan et al., 2008). Because the nucleocapsid protein is highly conserved, this has been successfully cloned and used to detect antibody response by enzyme immunoassay (EIA) and Western blot analysis of sera from infected human (Woo et al., 2004; Temperton et al., 2005). Common respiratory symptoms accompanying HKU1 infection are rhinorrhoea, fever, couging, wheezing, and disease manifestations include bronchiolitis and pneumonia, and it might also be involved in gastrointestinal disease (Lau et al., 2006; Sloots et al., 2006; Vabret et al., 2006; Sloots et al., 2008; Chan et al., 2009).

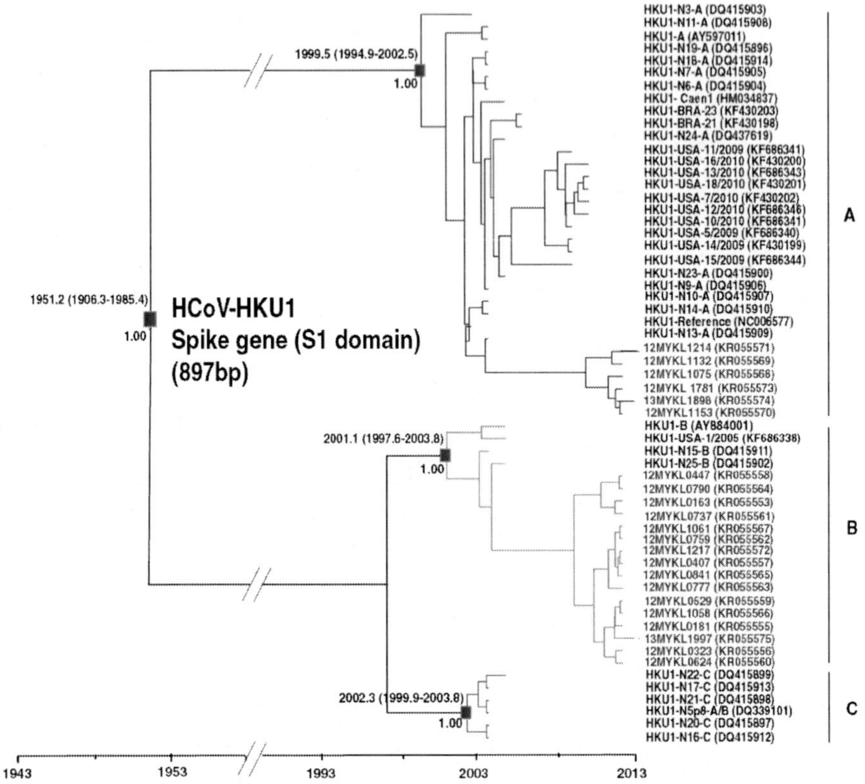

Figure 1. Maximum clade credibility (MCC) tree of HCoV-HKU1 genotypes. Estimation of the time of the most recent common ancestors (tMRCA) with 95% highest posterior density (95% HPD) of HCoV-HKU1 genotypes based on the spike gene (S1 domain) (897 bp). Data were analyzed under relaxed molecular clock with GTR + I substitution model and a constant size coalescent model implemented in BEAST. The Malaysian HCoV-HKU1 isolates obtained were color-coded and HCoV-HKU1 genotypes (a) to (c) were indicated. The MCC posterior probability values were showed on the nodes of each genotype (Al-Khannaq et al., 2016).

Recombination events in the spike (S), nucleocapsid (N) and the RNA dependent RNA polymerase (RdRp) within the 1a gene of HCoV-OC43 and HCoV-HKU1 leading to the emergence of unique recombinant genotypes have been found (Lau et al., 2011, Al-Khannaq et al., 2016). HCoV-HKU1 is a coronavirus that has been detected in adults with pneumonia in China, and in small groups of children with respiratory or enteric infections in China, Australia, France and the United States (Woo et al., 2005a,b,c; Esper et al., 2006; Lau et al., 2006; Sloots et al., 2006;

Vabret et al., 2006). Detecting HCoV-HKU1 in the nasopharyngeal secretions of a premature bronchiolitic infant confirms that HCoVs can cause moderate/severe respiratory infections (Fouchier et al., 2005; Kahn and McIntosh, 2005; Bosis et al., 2007). Esper et al., (2010) showed that HCoV-HKU1 can be identified in stool samples from children and adults with gastrointestinal disease with most individuals having respiratory findings as well. Jin et al., (2010) reported that HCoV-HKU1 is an uncommon virus existing among Chinese children with ARTI, and children with underlying diseases are more vulnerable to viral infection, and only HCoV-HKU1 genotype B circulated locally.

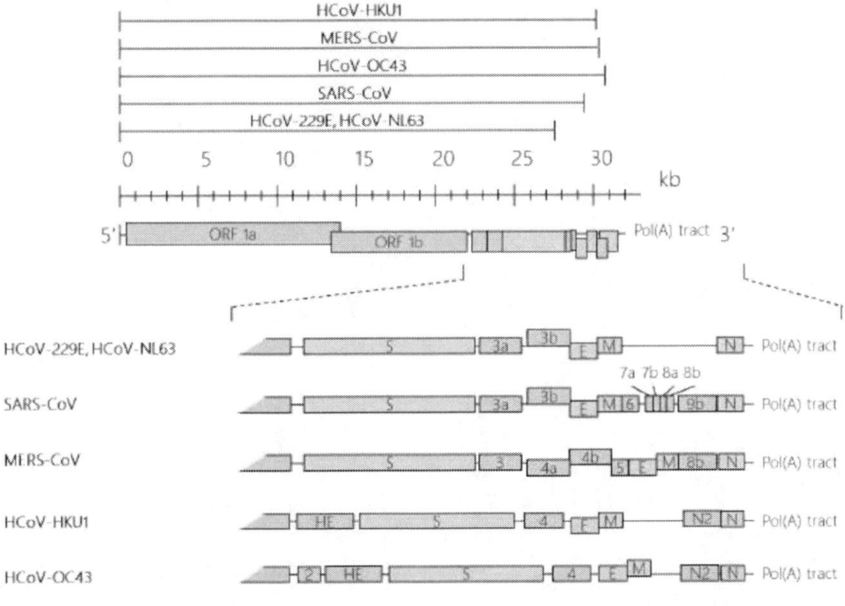

Figure 2. Genome organization of human coronaviruses (HCoVs). HCoV genomes range from about 26 to 32 kilobases (kb) in size, as indicated by the black lines above the scale. Coronavirus (CoV) genome is typically arranged in the order of 5′-ORF1a-ORF1b-S-E-M-N-3′. The overlapping open reading frames (ORF) ORF1 and ORF1b comprise two-thirds of the coronavirus genome, which encodes for all the viral components required for viral RNA synthesis. The other one-third of the genome at the 3′ end encodes for a set of structural (orange) and non-structural proteins (green) (Lim et al., 2016).

It has been reported that HKU1 and OC43 associated with benign common colds in healthy immunocompetent individuals, may cause significant morbidity and even mortality in the frail (Morfopoulou et al., 2016). Maximum clade credibility (MCC) tree of HCoV-HKU1 genotypes is shown in Figure 1. Genome organization of human coronaviruses (HCoVs) is presented in Figure 2.

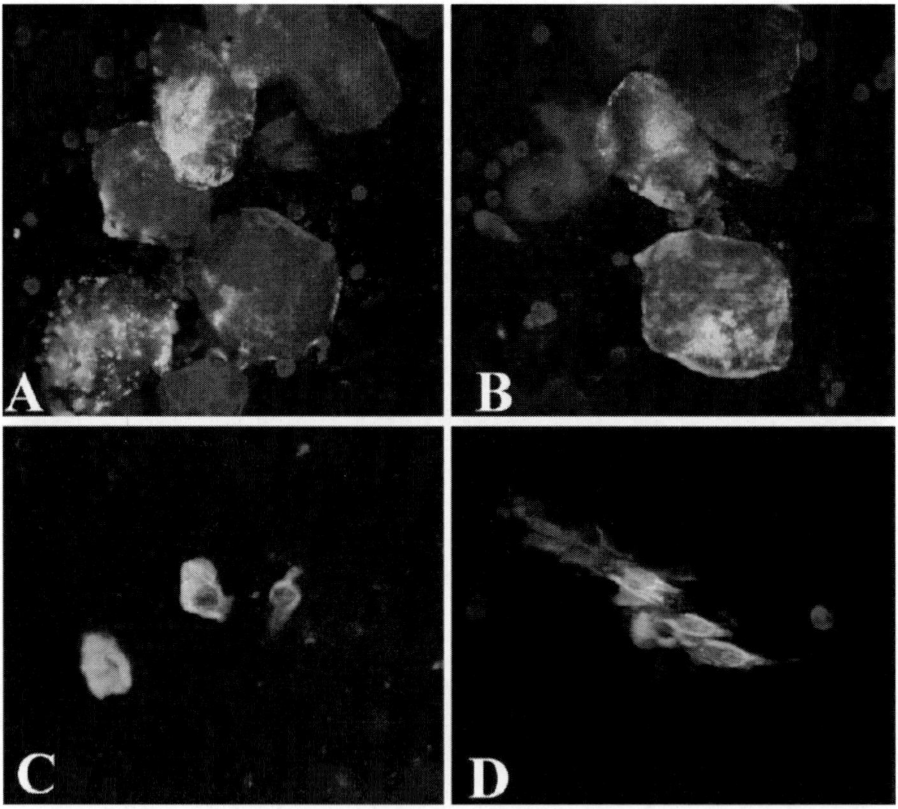

Figure 3. Detection of HKU1-positive respiratory cells from nasopharyngeal aspirates by direct staining using HKU1-specific monoclonal antibody. (A and B) Membrane staining of large syncytial formations. (C and D) Cytoplasmic staining of respiratory cells (Gerna et al., 2007).

The genome of HCoV-HKU1 is a 29,,926-nucleotide, polyadenylated RNA, the GC content is 32%, the lowest among all known coronaviruses, the genome organization is the same as that of other group II coronaviruses, with the characteristic gene order 1a, 1b, HE, S, E, M, and N (Pyrc et al., 2007). Moreover, accessory protein genes are present between the S and E gene (ORF4) and at the position of the N gene (ORF8). The TRS is presumably located within the AAUCUAAAC sequence, which precedes each ORF except E (Pyrc et al., 2007). Detection of HKU1-positive respiratory cells from nasopharyngeal aspirates by direct staining using HKU1-specific monoclonal antibody is indicated in Figure 3.

Chapter 4

Human Coronavirus 229E (HCoV-229E)

The 229E strain and the OC43 strain are only two HCoV were known before SARS epidemic in 2002-2003 (Geller et al., 2010). Four coronaviruses, HCoV-229E, HCoV-NL63, HCoV-OC43, and HCoV-HKU1, circulate in the human population where they are responsible for nearly one-third of the common cold (Li et al., 2019); HCoV-229E and HCoV-NL63 are closely related alphacoronaviruses that use different host proteins as receptors (Yeager et al., 1992; Hofmann et al., 2005). HCoV-229E uses human aminopeptidase N (hAPN) as receptor (Yeager et al., 1992), and it was recently shown in a cell-based entry assay that the camel 229E-like CoV can also use hAPN (Corman et al., 2016). Zhang et al. (2014) found that the HCoV-229E ORF4a protein is functionally analogous to the SARS-CoV 3a protein, which also acts as a viroporin that regulates virus production. The possibility that camels were involved in the transmission of HCoV-229E to humans is interesting given that the highly virulent MERS-CoV is directly transmitted to humans from dromedaries in many cases (Paden et al., 2018).

Its ectodomain is composed of the N-terminal S1 region that harbors the receptor binding domain (RBD) and the C-terminal S2 region that mediates membrane analysis has provided structures for both conformations (Shang et al., 2018a,b; Tortorici et al., 2019). HCoV-229E is complex responsible for receptor binding. Furthermore, those loops are the most variable region in the entire viral genome. HCoV-229E has been recognized as an important cause of Nosocomial Respiratory Viral Infections (NRVI) in high-risk infants (Sizun et al., 2001; Gagneur et al., 2002; Vallet et al., 2004). Hays and Myint (1998) reported that the translated spike protein sequence of human coronavirus 229E A162 showed three clusters of amino acid changes, situated within the 5′ half of the translated spike protein sequence. Yamaya et al., (2019) found that glycopyrronium, formoterol, and a combination of glycopyrronium, formoterol, and budesonide inhibit HCoV-229E replication partly by inhibiting receptor expression and endosomal function and that these drugs modulate infection-induced inflammation in the airway. Lo et al., (2013) stated that the C-terminal tail peptide is able to interfere with the oligomerization of the CTD of HCoV-229E N protein and performs the inhibitory effect on viral titre of HCoV-229E. Wang et al., (2014) reported that the production of human coronavirus 229E (HCoV-229E) progeny viruses, whose budding occurs at the ER-Golgi intermediate compartment (ERGIC), markedly decreases in the presence of BST2; and BST2 knockdown expression results in enhanced HCoV-229E virion production. They have suggested that BST2 exerts a broad blocking effect against enveloped virus release, regardless of whether budding occurs at the plasma membrane or intracellular compartments. Herold et al., (1993) concluded that the polymerase gene is comprised of two large open reading frames, ORF1a and ORF1b, which contain 4086 and 2687 codons, respectively. ORF1b overlaps ORF1a by 43 bases in the (-1) reading frame. The invitro translation of SP6 transcripts which include HCV 299E sequences encompassing the ORF1a/ORF1b junction showed that expression of ORF1b can mediated by ribosomal frame-shifting. The predicted translation products of ORF1a (454,200 molecular weight), and ORF1a/1b (754,200 molecular weight) have been compared to the predicted RNA polymerase gene products of infectious bronchitis virus (IBV) and murine hepatitis virus (MHV) and conserved structural features and putative functional domains have been identified.

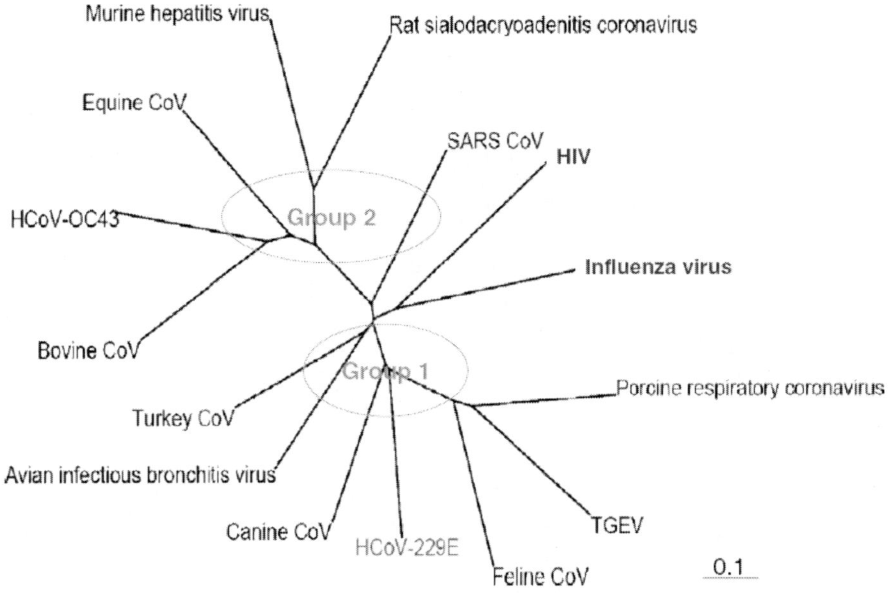

Figure 1. Phylogenetic tree of fusion proteins from some related coronaviruses. HCoV-229E is highlighted in red; HIV and influenza virus, two of the most classic coronaviruses, are indicated in blue; two different groups known are presented in green circles (Liu et al., 2006).

Desforges et al., (2007) suggested that infected monocytes could serve as a reservoir for HCoV-229E, become activated participate in the exacerbation of pulmonary pathologies, as well as serve as potential vectors for viral dissemination to host tissues, where it could be associated with other pathologies. Phylogenetic tree of fusion proteins from some related coronaviruses is shown in Figure 1.

Resolution of the HCoV-229E S-protein cryo-EM map is indicated in Figure 2. Key hydrophobic interactions defining assembly of the HCoV-229E S-protein trimer is shown in Figure 3. Schematic representation of the genomic organization of group 1 coronaviruses (CoVs) is presented in Figure 4.

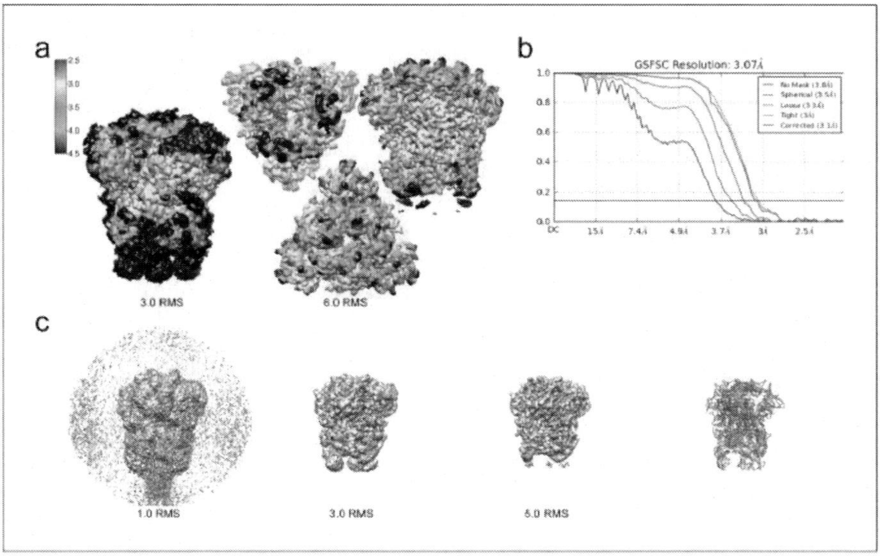

Figure 2. Resolution of the HCoV-229E S-protein cryo-EM map. (a) Local resolution (A) plotted on the cryo-EM map surface as a heat map. Leftmost image, a side view of the map contoured at 3.0 RMS to show the weak densities of the connector domains (at bottom). To the right, three views of the map contoured at 6.0 RMS showing that the entire S1 region and the helical core region of S2 are well-resolved. (b) GSFSC curves of the final 3D non-uniform refinement from cyroSPARC v2. The blue horizontal line indicates an FSC value of 0.143. (c) The cryo-EM map contoured at different levels to show the weak densities near the C-terminal end (bottom). To the right, the atomic model is shown in ribbon representation with the well resolved regions of the S1 and S2 subunits colored cyan and magenta, respectively. The poorly resolved connector domains are colored gray (Li et al., 2019).

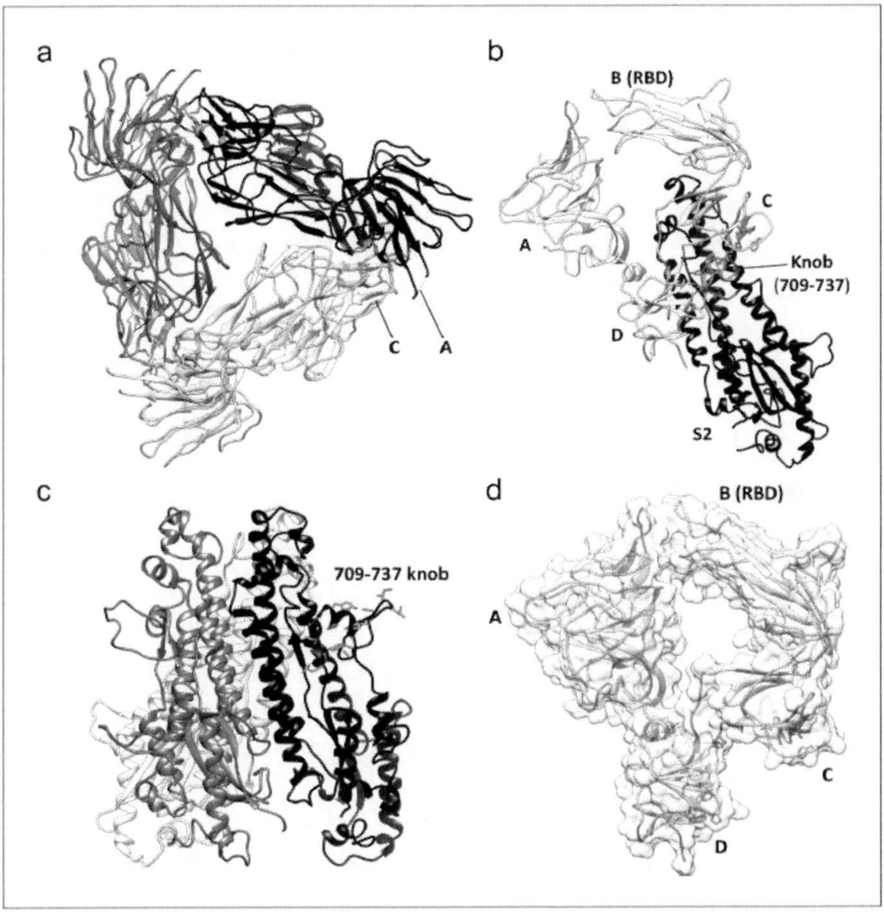

Figure 3. Key hydrophobic interactions defining assembly of the HCoV-229E S-protein trimer. (a) Hydrophobic interactions in the S1 region. Each monomer is colored white, gray and black, respectively. Apolar residues between the S1 subunit are colored cyan. (b) The S1 subunit of one monomer (white) interacts with the S2 subunit of another monomer (black). The C and D domains of the S1 subunit form a hydrophobic clamp over the knob formed by S2 subunit residues 709-737 (magenta); interacting apolar side chains are shown in cyan. (c) Side view of the S2 region of the trimer showing the 709-737 knob; apolar residues are colored ccyan (on one monomer only). (d) The hydrophobic clamp is formed by domains C and D; apolar residues in contact with the 709-737 knob are shown in cyan (Li et al., 2019).

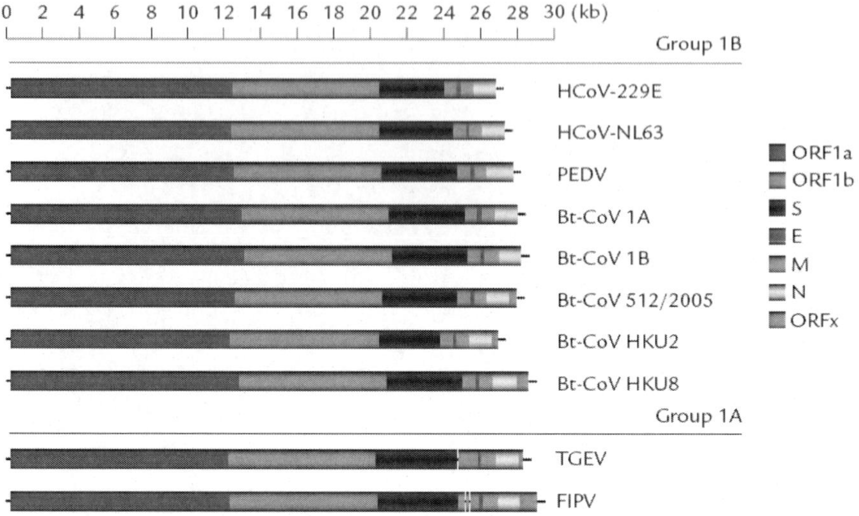

Figure 4. Schematic representation of the genomic organization of group 1 coronaviruses (CoVs). Group 1B CoVs HCoV-229E (NC002645), HCoV-NL63 (NC005831), porcine epidemic diarrhea virus (PEDV: NC003436), bat coronavirus 1A (Bt-CoV 1A: NC010437), Bt-CoV 1B (NC010436), Bt-CoV 512/2005 (NCoo9657), Bt-CoV HKU2 (NC009988) and Bt-CoV HKU8 (NC010438), and group 1A CoVs porcine transmissible gastroenteritis virus (TGEV: NC002306) and feline infectious peritonitis virus (FIPV: NC007025) genome organization. The open reading frames (ORFs) are denoted as replicase 1A (ORF1a), replicase 1B (ORF1b), S, E, M, N and accessory genes (ORFx) (Dijkman and van der Hoek, 2009).

Schematic representation of the amplification products generated using the human coronavirus 229E spike gene nested PCR is shown in Figure 5. Downregulation of Cyp B, but not CypA in the presence of Cyp inhibitors CsA and ALV is indicated in Figure 6. Homology modeling of HCoV-229E S protein is presented in Figure 7. Structure modeling of HCoV-229E fusion core is shown in Figure 8.

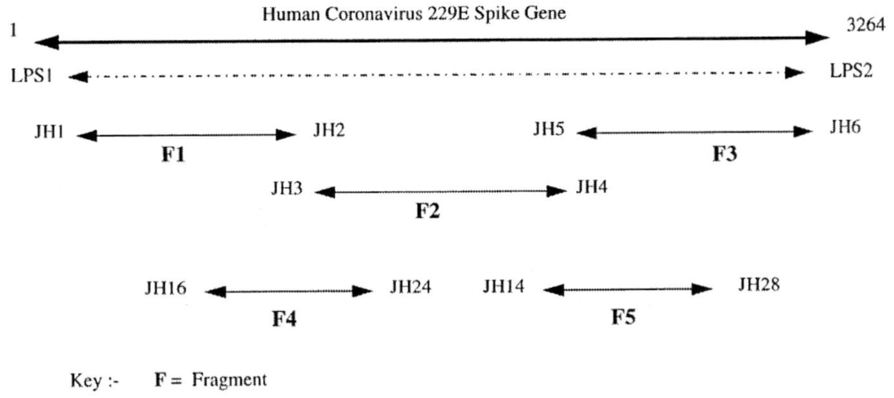

Figure 5. Schematic representation of the amplification products generated using the human coronavirus 229E spike gene nested PCR (Hays and Myint, 1998).

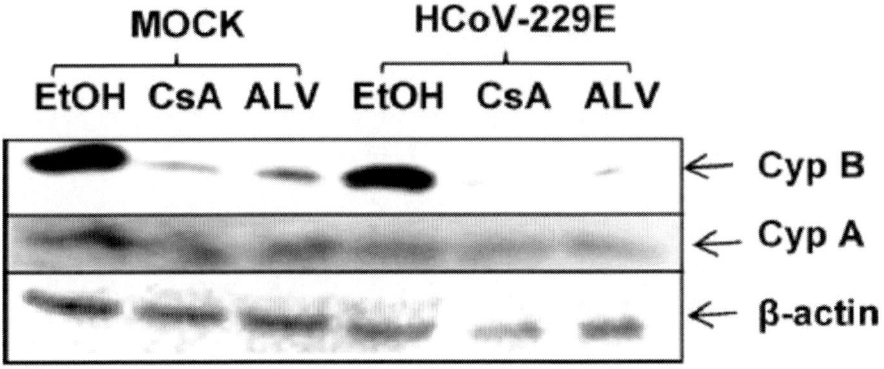

Figure 6. Downregulation of Cyp B, but not CypA in the presence of Cyp inhibitors CsA and ALV. Huh7 cells were either mock or HCoV-229E (MOI = 1) infected and cultivated in the presence of EtOH solvent or 20 μM CsA or LAV for 48 h. Cell extracts were subjected to Western blot analysis and staining with anti-Cyp A, anti-Cyp B and anti-beta actin as loading control (Ma-Lauer et al., 2020).

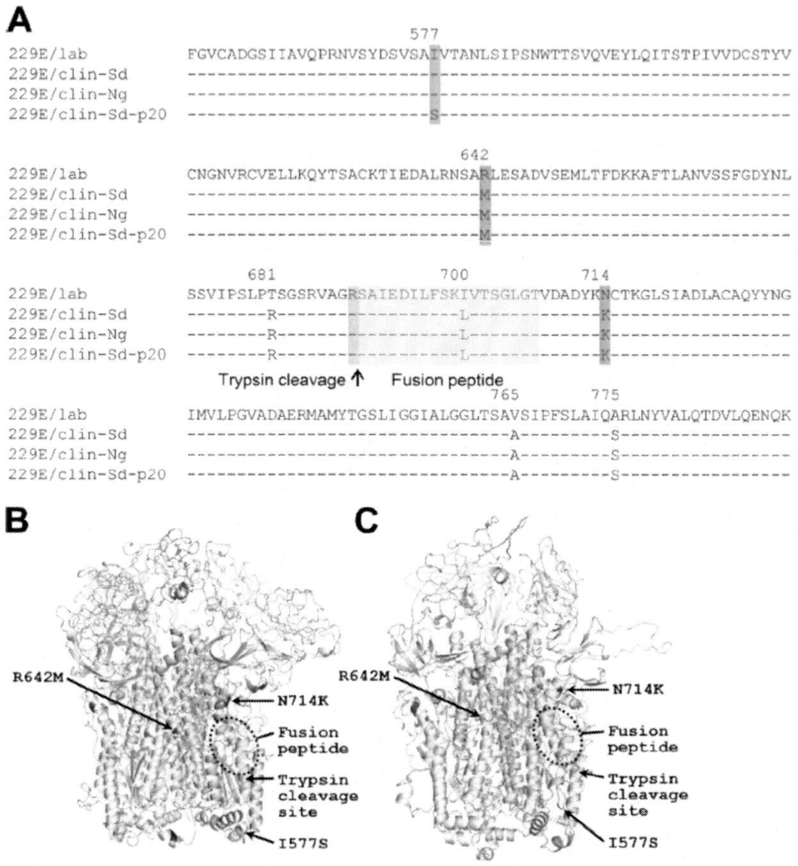

Figure 7. Homology modeling of HCoV-229E S protein. (A) Alignment of HCoV-229E S proteins around a fusion peptide. The intermediate regions between the S1 and S2 subunits of the HCoV-229E S glycoproteins were aligned using MAFFT software (CBRC, Japan). The trypsin cleavage site and the fusion peptide are indicated in green and yellow, respectively. The positions of the mutations causing an increase or decrease in cathepsin use are indicated in red or blue, respectively. (B and C) Theoretical structures of the S protein. The structures of HCoV-229E were constructed using MOE software based on the cryoelectron microscopic structure of the MHV S protein (B) or the HCoV-HKU1 S protein (C) in the prefusion state. The region around the fusion peptide in the S2 subunit (resides 572 to 953) is shown in orange. Mutations that cause increased or decreased cathepsin usage are shown in blue or red, respectively. The putative fusion peptide and the trypsin cleavage site are shown in yellow and green (Shirato et al., 2017).

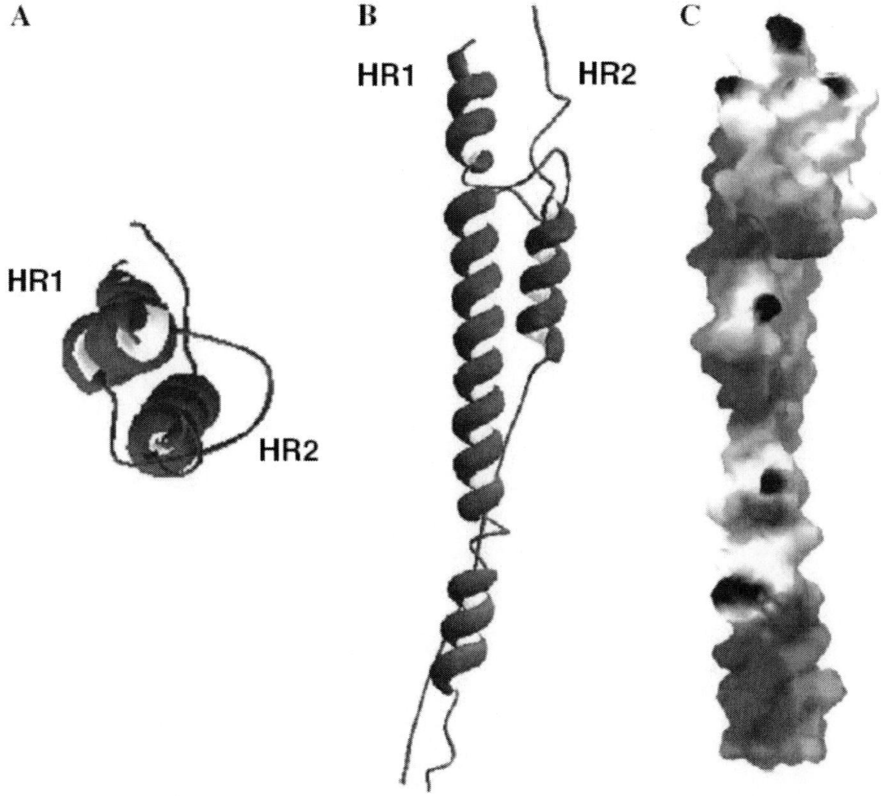

Figure 8. Structure modeling of HCoV-229E fusion core. (A) Top view of the core structure in ribbon. (B) Front view of the core structure in ribbon. (C) Surface view of the core structure (blue, +; organge, -; others, low charge) (Liu et al., 2006).

Chapter 5

HUMAN CORONAVIRUS NL63 (HCoV-NL63)(NEW HAVEN CORONAVIRUS)

The human coronavirus NL63 (HCoV-NL63) was first identified in the Netherlands, and its circulation in France has not been investigated (Pyrc et al., 2006). It was already isolated in 1988 or even earlier (Fouchier et al., 2004; van der Hoek et al., 2006; Banach et al., 2009). NL63 infection generally caused mild upper respiratory tract disease, but may also cause more severe lower respiratory tract diseases, e.g., croup, bronchiolitis, and pneumonia in young children, the elderly and immunocompromised people (van der Hoek et al., 2006). No vaccine or antiviral drug is currently available for NL63 (Lin et al., 2011). Cui et al. (2011) reported that in HCoV-NL63-infected patients they were cough, fever, and rhinorrhea. Han et al., (2007) found that HCoV-NL63 may be one of the causative agents of acute respiratory tract infection, especially croup. An important aspect of HCoV-NL63 infection is the co-infection with other human coronaviruses, influenza A, respiratory syncytial virus (RSV), parainfluenza virus or human metapneumovirus (Abdul-Rasool and Fielding, 2010; Carbajo-Lozoya et al., 2014). Van der Hoek et al., (2010) found that HCoV-NL63 infection in children below 3 years of age often requires a visit to the physician in an outpatient clinic, especially during peak-year, but hospitalizations are relatively infrequent. Dijkman et al.,

(2012) showed that HCoV-NL63 and HCoV-OC43 infections occur frequently in early childhood, more often than HCoV-HKU1 or HCoV-229E infection. HCoV-OC43 and HCoV-Nl63 may elicit immunity that protects from subsequent HCoV-HKU1 and HCoV-229E infection, respectively, which would explain why HCoV-OC43 and HCoV-NL63 are the most frequent infecting HCoVs. Schematic diagram of genome organization of NL63 and domain architecture spike (S) protein is shown in Figure 1. Comparison of clinical symptoms of human coronaviruses is presented in Table 1. Genome organization of HCoV-NL63 and other group I coronaviruses is shown in Figure 2. The colored regions of each human coronavirus indicate genetic recombination at the genomic sites with coronaviruses of zoonotic origin is presented in Figure 3. Structural overview of HCoV-NL63 M^{pro} is presented in Figure 4. S1 and S2 binding sites in HCoV-NL63 M^{pro} is shown in Figure 5. The SARS-CoV RBD inhibits HCoV-NL63 S-protein-mediated infection I shown in Figure 6.

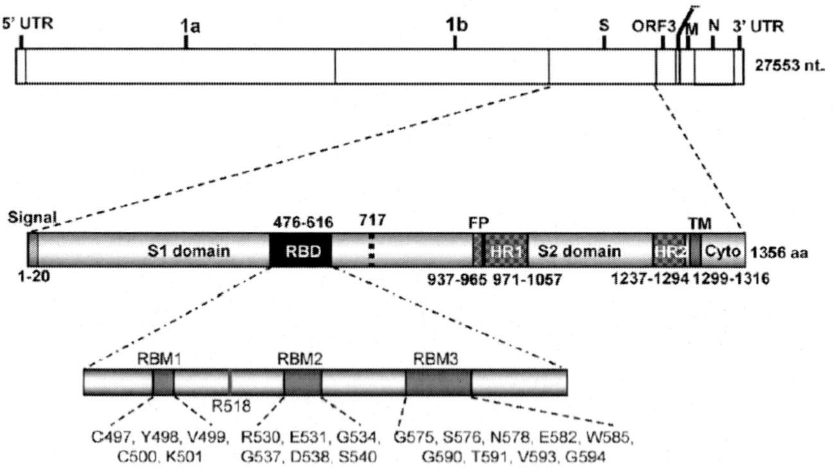

Figure 1. Schematic diagram of genome organization of NL63 and domain architecture spike (S) protein. Signal: signal peptide; RBD: receptor-binding domain; FP: fusion peptide; HR: heptad repeat; TM: transmembrane domain; Cyto: cytoplasmic tail. Twenty-one residues that have been previously shown to be important for RBD-hACE2 interaction are indicated within three receptor binding motifs (RBMs) (Li et al., 2007; Lin et al., 2008).

Table 1. Comparison of clinical symptoms of human coronaviruses

Human coronaviruses	Clinical Symptoms
229E	General malaise Headache Nasal discharge Sneezing Sore throat Fever and cough (10-20% of patients)
OC43	General malaise Headache Nasal discharge Sneezing Dore throat Fever and cough (10-20% of patients)
SARS-CoV	Fever Myalgia Headache Malaise Chills Nonproductive cough Dyspnea Respiratory distress Diarrhea (30-40% of patients)
NL63	Cough Rhinorrhea Tachypnea Fever Hypoxia Obstructive laryngitis (croup)

Table 1. (Continued)

Human coronaviruses	Clinical Symptoms
HKU1	Fever Running nose Cough Dyspnea
MERS-CoV	Fever Cough Chills Sore throat Myalgia Arthralgia Dyspnea Pneumonia Diarrhea and vomiting (one-third of patients) Acute renal impairment

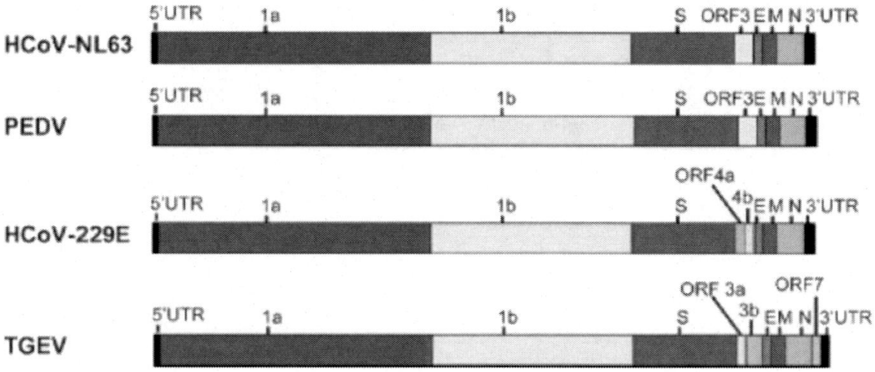

Figure 2. Genome organization of HCoV-NL63 and other group I coronaviruses (van der Hoek et al., 2006).

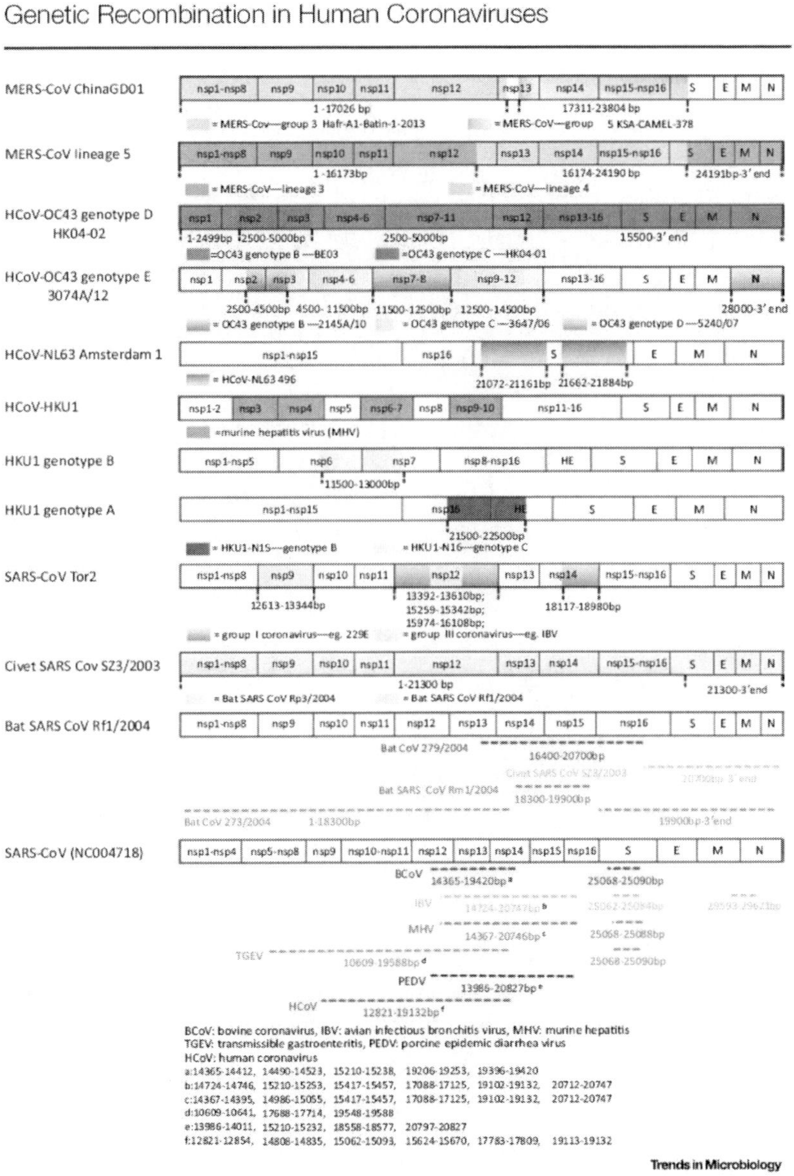

Figure 3. The colored regions of each human coronavirus indicate genetic recombination at the genomic sites with coronaviruses of zoonotic origin. Specific nucleotide locations where recombination occurred are also shown in broken lines (Su et al., 2016).

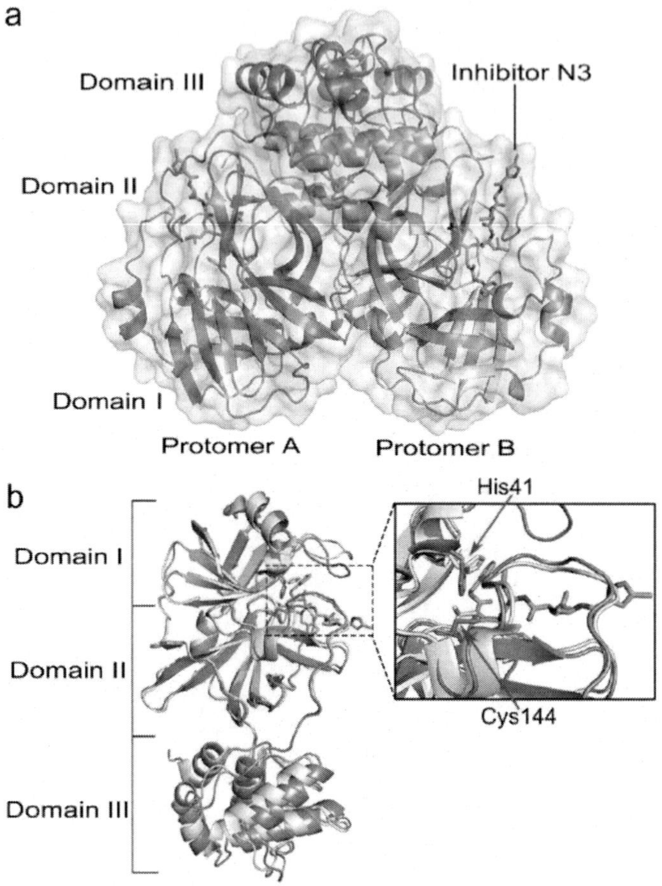

Figure 4. Structural overview of HCoV-NL63 M^{pro}. (a) Overview of homodimer in one asymmetric unit (A: slate and B: deep salmon). Protomers are shown in cartoons, and N3 inhibitors are shown as green sticks. (b) Structural alignment of protomer A in M^{pro}-N3 (slate) complex with that in apo enzyme (light orange, PDB ID: 3TLO). The backbone is shown in cartoons, and N3 inhibitor is presented as green sticks (Wang et al., 2016).

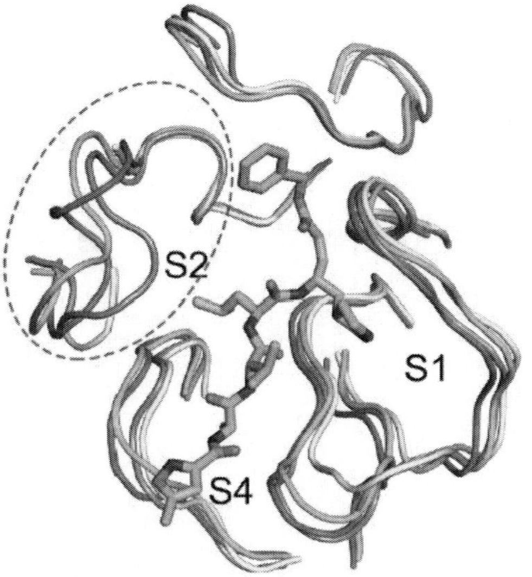

Figure 5. S1 and S2 binding sites in HCoV-NL63 M^pro^. The main chains of four human CoV M^pro^ structures (HCoV-NL63: slate, HCoV-229E: cyan, SARS-CoV: magenta, and HCoV-HKU1: yellow) are superimposed and displayed in the neighborhood of the substrate-binding site. The S1, S2 and S4 binding sites are labeled. The backbones are represented in worm form, an inhibitor is shown in stick format of green color. Residues 45-51 are marked with a red oval (in dash line) (Wang et al., 2016).

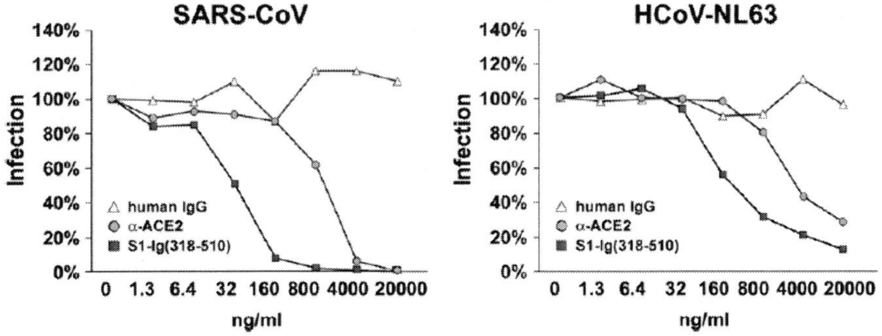

Figure 6. The SARS-CoV RBD inhibits HCoV-NL63 S-protein-mediated infection (Li et al., 2007). Lentiviruses expressing luciferase and pseudotyped with the S proteins of SARS-CoV or HCoV-NL63 were incubated with ACE2-expressing HEK293T cells in the presence of indicated concentrations of affinity purified goat anti-ACE2 antibody, SARS-CoV RBD-Ig, or human IgG. Infection is expressed as a percentage of luciferase activity observed in the absence of inhibitor.

Chapter 6

FROM SARS TO COVID-19 (SARS, MERS AND SARS-COV-2)

Virus is a tiny agent, around one-hundredth the size of a bacterium which can infect cells of plants and animals. A coronavirus has spikes around its spherical body (corona = crown). Coronaviruses are enveloped, positive-sense, single-stranded RNA viruses of the family Coronaviridae (Huang et al., 2007; Kam et al., 2007; Chu et al., 2008; Masters and Perlman, 2013). Corona virus was first identified in the 1960s and is recognized causes of mild respiratory tract infections in humans (Kumaki et al., 2011; Bermingham et al., 2012; Karypidou et al., 2018). The first two HCoVs, HCoV-229E and HCoV-OC43 have been known since the 1960s (Chang et al., 2020). The viruses are subdivided into four genera on the basis of genotypic and serological characters which are Alpha-, Beta-, Gamma, and Deltacoronavirus (Akerstrom et al., 2009; Adams and Carstens, 2012), and among them the first two genera are those which infect humans (Sims et al., 2008; Lu et al., 2015; Ar Gouilh et al., 2018). Seven coronaviruses are known to infect humans, three of them are serious, namely, SARS (severe acute respiratory syndrome, China, 2002), MERS (Middle East respiratory syndrome, Saudi Arabia, 2012), and SARS-CoV-2 (2019-2020). SARS-CoV, and MERS-CoV belong to betacoronaviruses (betaCoVs) (Lu et al., 2015).

Table 1. History of epidemics since 1900

Year	Outbreak
1918	Great flu pandemic
1976	Legionnaires disease
1993	Hanta virus pulmonary syndrome
1994	Hendra virus infection
1997	H5N1 influenza infection
1999	Nipah virus encephalitis/pneumonitis
2002	Severe acute respiratory syndrome (SARS)
2012	Middle east respiratory syndrome (MERS)
2019	Severe acute respiratory syndrome coronavirus 2 (SARS-CoV-2)

Table 2. Three main coronavirus outbreaks and their origin

Coronavirus outbreak	Origin
SARS	Civets (although, the virus originated in bats, and civets consider as intermediary)
MERS	Camels (The virus also came from bats)
SARS-CoV-2	Malayan pangolins (armadillo-like mammals)(The virus also came from bats)

Table 3. Important steps allow the virus entry (Millet and Whittaker, 2018)

a) Bind to a target host cell, typically via interactions with cellular receptors.
b) Fuse its envelope with a cellular membrane, either at the plasma membrane or through the endocytic pathway.
c) Deliver its genetic material inside the cell.

For enveloped viruses, a critical player in the entry process is the viral fusion protein as it mediates the membrane fusion reaction (Colman and Lawrence, 2003; Chernomordik and Kozlov, 2008; Harrison, 2008; White et al., 2008; White and Whittaker, 2016). The viral entry into target cells for coronaviruses is performed by the spike (S) envelope glycoprotein, which mediates both host cell receptor binding and membrane fusion (Bosch et al., 2003; Millet and Whittaker, 2018). History of epidemics diseases since 1900 has shown in Table 1. Three main coronavirus outbreaks and their origin indicated in Table 2. Important steps allow the virus entry is shown in Table 3.

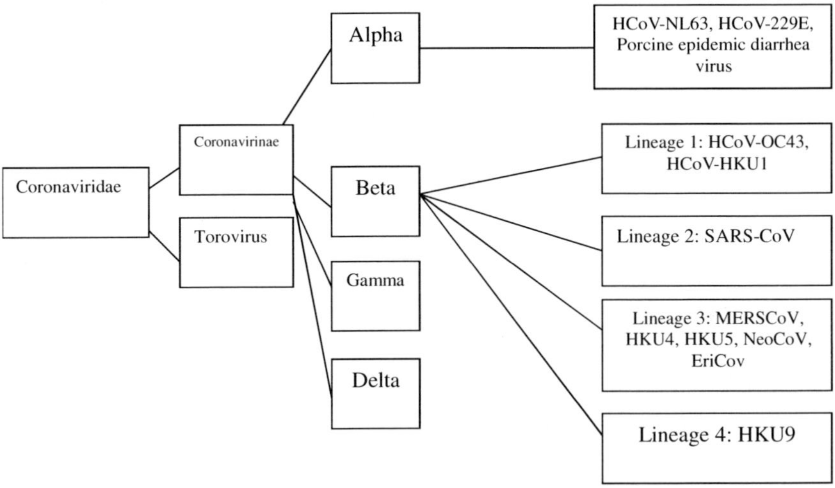

Figure 1. Taxonomy of the Coronaviridae family.

Table 4. Host information and distribution of SARSr-CoVs available in GenBank (Yu et al., 2019)

Provinces	**Bat species**
Guangdong	*Rhinolophus sinicus*
Guangxi	*Rhinolophus pearsonii, Rhinolophus sinicus*
Guizhou	*Rhinolophus rex, Rhinolophus sinicus*
Hebei	*Rhinolophus ferrumequinum*
Henan	*Rhinolophus ferrumequinum*
Hong Kong	*Rhinolophus sinicus*
Hubei	*Rhinolophus ferrumequinum, Rhinolophus macrotis, Rhinolophus sinicus*
Jilin	*Rhinolophus ferrumequinum*
Shaanxi	*Rhinolophus pusillus*
Shanxi	*Rhinolophus ferrumequinum*
Taiwan	*Rhinolophus monoceros*
Yunnan	*Aselliscus stoliczkanus, Rhinolophus affinis, Rhinolophus ferrumequinum, Rhinolophus sinicus*
Zhejiang	*Rhinolophus monoceros, Rhinolophus pearsonii, Rhinolophus sinicus, Rhinolophus thomasi*

Taxonomy of the Coronaviridae family is presented in Figure 1. Host information and distribution of SARSr-CoVs available in GenBank is presented in Table 4. The aim of this manuscript is review of SARS, MERS and COVID-19.

Chapter 7

SEVERE ACUTE RESPIRATORY SYNDROME (SARS)

The contagious and sometimes fatal severe acute respiratory syndrome (SARS) is a respiratory illness which was first appeared in China in 2002, and it did spread worldwide, mostly by unsuspecting travelers (Gillissen and Ruf, 2003; Zhong and Wong, 2004; Chu et al., 2005). It is caused by a coronavirus, which was called SARS coronavirus (SARS-CoV or SARS CoV-1). It is similar to other coronaviruses in both virion structure and genome organization with a singlestranded, plus-sense RNA (Zhai et al., 2007; Hong et al., 2009). CoVs are single-stranded RNA viruses which belong to the order Nidovirales, family Coronaviridae, and subfamily Coronavirinae (Yam et al., 2007; Zheng et al., 2015; Schoeman and Fielding, 2019), and have been classified into four major groups: α-CoVs, β-CoVs, γ-CoVs, and δ-CoVs with 17 sub-types (Saminathan et al., 2014). It has been reported, SARS-CoV, like other coronaviruses, is an RNA virus which replicates in the cytoplasm, and the virion envelope contains at least three structural proteins, S, E, and M, embedded in the membrane, and also like other coronaviruses, SARS-CoV encodes several group-specfici proteins, termed 3a, 3b, 6, 7a, 7b, 8 and 9 b (Snijder et al., 2003; Stockman et al., 2006; Netland et al., 2010). Deletion of the small envelope (E) protein modestly reduces SARS-CoV growth *in vitro* and *in vivo* (DeDiego et al., 2007, 2008; Ohnishi et al., 2019), which may result in an attenuated virus. SARS 8b, known as X5 is predicted to be a soluble protein with 84 amino acids and an estimated size of 9.6 kDa (Law et al., 2006). It showed minor homology to the human coronavirus E2glycoprotein precursor (Rota et al., 2003). Other researches also showed that aside from four typical structural proteins, nucleocapsid (N), envelope (E), membrane (M), and spike (S) protein and nearly 16 non-structural proteins (Nsp1-16) involved in viral replication, SARS-CoV encodes an exceptionally high number of accessory proteins that bear little resemblance to accessory genes of other coronaviruses (Shin et al., 2006; Narayanan et al., 2008; Nguyen et al., 2011; Liu et al., 2014; Schoeman and Fielding, 2019). Like other coronaviruses, SARS-CoV is an inefficient inducer of IFN-β response in cell culture system (Spiegel et al., 2005) and is sensitive to the antiviral state induced by IFNs (Spiegel et al., 2004; Zheng et al., 2004). Two functional domains of S protein, S1 and S2 located in the N- and C-terminal regions, respectively, of the S protein are conserved among the coronaviruses (He et al., 2004). The S protein in coronaviruses are major antigenic determinants that induce immune response in the hosts (Gallangher and Buchmeier, 2001; Holmes, 2003a,b). The S protein of transmissible gastroenteritis virus contains four major antigenic sites (A to D), and site A on the S1 subunit is the main inducer of neutralizing Abs (Gebauer et al., 1991; Posthumus et al., 1991; Enjuanes et al., 1992). Ab responses to SARS-CoV can be developed in SARS patients; but, its antigenic determinants remain to be elucidated (Li et al., 2003). But, in other coronaviruses, deletion of E results in either complete absence of infectious virus or a severe reduction in titer (Kuo and Masters, 2003; Ortego et al., 2007). DeDiego et al., (2014) reported that E protein is responsible in a significant proportion of the inflammasome activation and the associated inflammation elicited by SARS-CoV in the lung parenchyma, and the inflammation may lead to edema accumulation which cause acute respiratory distress syndrome (ARDS). E protein contains several active motifs despite its small size, between 76 and 109 amino acids depending on the CoV (DeDiego et al., 2011; Nieto-Torres et al., 2014). Schematic representation of the taxonomy of *Coronaviridae* is shown in Figure 1. The most important characteristic of SARS is shown in Table 1. Structure of SARS-CoV E protein proteolipidic ion channel is indicated in Figure 2.

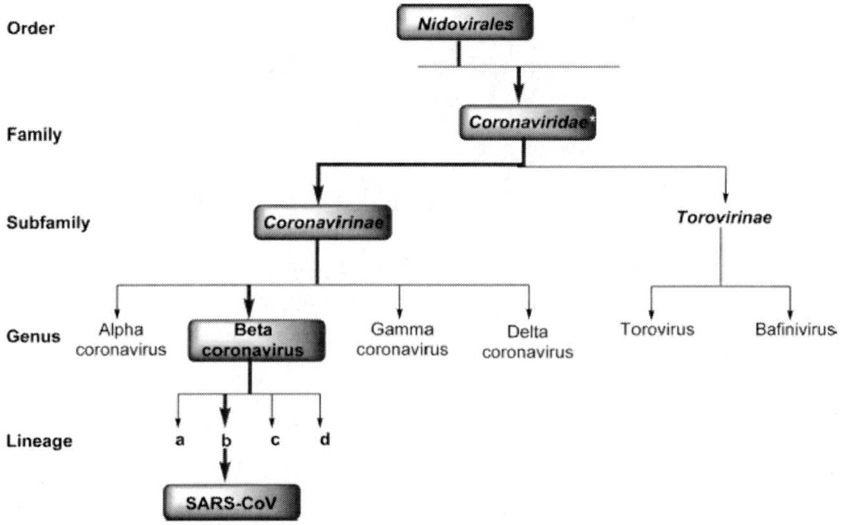

Figure 1. Schematic representation of the taxonomy of *Coronaviridae* (on the basis of International Committee on Taxonomy of Viruses). SARS-CoV belongs to the Betacoronavirus family but has a "b" lineage. Other members of this family are *Arteriviridae*, *Mesoniviridae*, and *Roniviridae* (Pillaiyar et al., 2016).

Table 1. The most important characteristic of SARS

a) The first international disease epidemic of the twenty-first century.
b) It is the first time this Coronavirus has been found in humans.
c) It was the first public health challenge for international communities.
d) The first disease that globalization visibly exacerbated its spread in a short time and contributed significantly to its end.

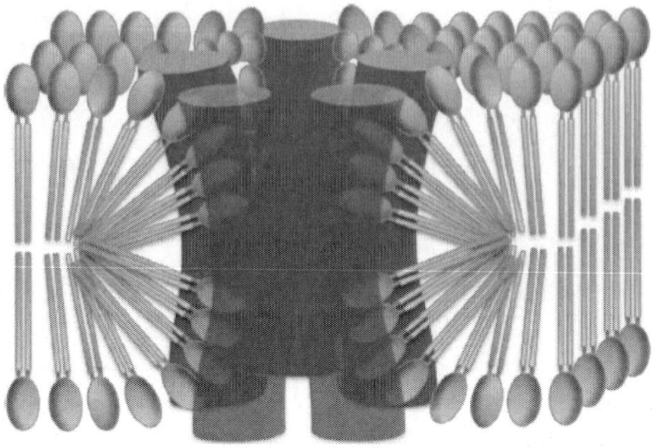

Figure 2. Structure of SARS-CoV E protein proteolipidic ion channel. Phospholipids are represented in blue, and E protein monomers are shown as red cylinders. Lipid head groupd (blue ellipses) also face the ion channel lumen (DeDiego et al., 2014).

SARS should not be confused with avian flu which is another zoonosis from the same area. It initially began in the Guangdong province in south of China in 2002-2003 which eventually involved more than 8400 people worldwide, which is around 9.5% of the total affected (Chee, 2003; Kisson, 2003; Thomas, 2003). The greatest number of SARS cases were in mainland China, Hong Kong, Taiwan, Singapore, Canada, respectively (Chowell et al., 2003; Lang et al., 2003; Ruiz-Contreras, 2003). Quick infection is one of the main character of SARS (Hawkey et al., 2003; Stohr, 2003; Hoheisel et al., 2007). The most important symptoms of SARS were fever, chills, muscle aches, headache and diarrhea, which may lead to fever with body temperature of 38 °C or higher, dry cough and shortness of breath after around one week (Patrick, 2003; Shin et al., 2007; Hui and Zumla, 2019). On the basis of former reports the features of the clinical examination found in the patients at admission were self-reported fever (99%), documented elevated temperature (85%), nonproductive cough (69%), myalgia (49%), and dyspnea (42%) (Booth et al., 2003; Chan-Yeung and Yu, 2003; Leung et al., 2003; Zhong, 2003). SARS spread through droplets which enter the air with coughs, sneezes or talks; moreover, it may spread on contaminated objects and surfaces like doorknobs, elevator buttons and telephones.

Table 2. The most important symptoms of SARS

Fever more than 30oC
Dry cough
Sore throat
Problems in breathing, such as shortness of breath, inability to maintain oxygenation (hypoxia)
Headache
Body aches and muscles pain
Loss of appetite
Malaise
Night sweats and chills
Confusion
Rash
Nausea, vomiting and diarrhea
Weakness
Fever
Poor appetite
Respiratory distress syndrome (ARD or ARDS)
Attacking the alveoli (air sacs) in the lungs
Kidney failure
Inflammation of the heart sac (pericarditis)
Sever systematic bleeding from disruption of clotting system (disseminated intravascular coagulation)
Reduced lymphocyte cell counts (lymphopenia)
Inflammation of the arteries (vasculitis)

SARS-CoV is transmitted mainly person-to-person infection. During its outbreak, nearly 25% of people had severe respiratory failure and 10% died, and it was controlled by using public-health measures, namely, wearing surgical masks, washing hands and isolating infected patients (Aronin and Sadigh, 2004; Shen et al., 2005; Lien et al., 2008; Ahmad et al., 2009). Face to face contact of SARS can be divided into three groups, 1) caring for someone with SARS, 2) having contact with the bodily fluids of a person with SARS, 3) Kissing, hugging, touching or sharing eating or drinking utensils with an infected person (Wang and Ruan, 2004; Derrick and Gomersall, 2005; Bryce et al., 2008). Almost 25 percentage of case developed severe pulmonary disease which may lead to death from respiratory failure (Tsui, 2003). SARS which had flu-like signs can lead to death in severe conditions due to respiratory failure or complications consist of heart and liver failure, especially for those old people who had diabetes and hepatitis (Tong et al., 2006; Wang et al., 2007; Wong et al., 2012). Like common cold, it is caused by a strain of corona virus. Coronaviruses may lead to sever disease in animals, and it is supposed that the SARS virus might have come from animals to humans (Poon et al., 2004; Li et al., 2007; Hafeez et al., 2016). SARS-CoV originated in wild bats and then spread to palm civets or similar mammals. Bats and Civets which are cat-like serve as foods and in folk medicines. Musk production from the scent glands of civets which is used in perfumes is another usage of civets. These mentioned animals could easily transmit the virus to humans (Wang et al., 2006). The most important symproms of SARS is shown in Table 2. Criteria for disease control and prevention case definition of SARS are shown in Table 3. The most important guidelines to avoid SARS infections are indicated in Table 4. Case classification of SARS is indicated in Table 5. The most important recommendations for health care workers are shown in Table 6.

Table 3. Criteria for disease control and prevention case definition of SARS (Centers for disease control and prevention, 2001)

Clinical Criteria	Characteristics
Asymptomatic or mild respiratory illness	
Moderate respiratory illness	Temperature of > 100.4o F(>38o C)
	One or more clinical findings of respiratory illness (eg, cough, shortness of breath, difficulty breathing, or hypoxia)
Severe respiratory illness	Temperature of >100.4o F (>38o)
	One or more clinical findings of respiratory illness (eg, cough, shortness of breath, difficulty breathing, or hypoxia)
	Radiographic evidence of pneumonia
	Respiratory distress syndrome
	Autopsy findings consistent with pneumonia or respiratory distress syndrome without an identifiable cause
Epidemiologic Criteria	Travel (including transit in an airport) within 10 days onset of symptoms to an area with current or previously documented or suspected community transmission of SARS

Clinical Criteria	**Characteristics**
	Close contact within 10 days of onset of symptoms with a person known or suspected to have SARS
Laboratory Criteria	
Confirmed	Detection of antibody to SARS-CoV in specimens obtained during acute illness or >21 days after illness onset
	Detection of SARS-CoV RNA by RT-PCR confirmed by a second PCR assay, by using a second aliquot of the specimen and a different set of PCR primers
	Isolation of SARS-CoV
Negative	Absence of antibody to SARS-CoV in convalescent serum obtained >21 days after symptom onset
Undetermined	Laboratory testing either not performed or incomplete

Table 4. The most important guidelines to avoid SARS infections

Wash hands, frequently with soap and water or alcohol-based hand rub.
Wear disposable gloves
Wear a surgical mask or eyeglasses
Wash personal items
Disinfect surfaces
Cover nose and mouth when sneeze or cough
Avoid contact with people who are coughing or sneezing
Stay at home and rest, is a person has its symptoms

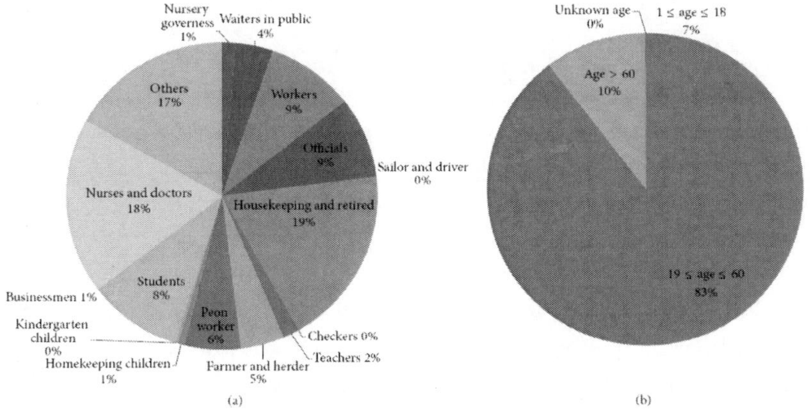

Figure 3. Basic patient information. (a) The occupation distribution as a percentage of all SARS infections. (b) The age percentage of SARS infections (Cao et al., 2016).

Table 5. Case classification of SARS

Under investigation	**A person who has been referred to the public health service for possible SARS-CoV infection**
Suspected case	A person with all of these following: a) Sudden onset of high fever, > 38oC b) One or more of the following respiratory symptoms: cough, sore throat, shortness of breath, and difficulty in breathing c) Showing symptoms within 10 days of either travelling to one of the suspected areas of SARS or being in close contact with a person who has travelled to those areas
Probable case	a) A suspected case with chest X-ray findings of pneumonia or adult respiratory distress syndrome b) A person with an unexplained respiratory illness resulting in death, with a post-mortem examination demonstrating the pathology or respiratory distress syndrome without an identifiable cause
Confirmed case	A clinically compatible illness which is confirmed by laboratories
Not a case	A case that has been investigated and subsequently found not to meet the case definition

Table 6. The most important recommendation for health care workers

Airborne precautions (such as isolation room, use filtering masks)
Contact precautions
Considering standard or universal precautions such are hand hygiene
Wear eye protection for all patient contact
Standard precautions when handling any clinical waste
Laundry should be categorized as infected
Considering World Health Organization (WHO), it hospitals or clinics lack isolation facilities and many cases occur
Long sleeve fluid repellent gowns
For environmental decontamination of areas, hypochlorite is the recommended disinfectant

No medication has been proven to treat SARS effectively, but oxygen therapy and tracheal intubation and mechanical ventilation to support life until recovery begins is useful for patients in severe cases (Woodhead et al., 2003). The most useful ways to control SARS pandemic is public-health and infection-control measures (Dolan, 2003; Wenzel et al., 2005). Peiris (2003) confirmed that its rapid mobilization and coordination of relevant expertise when it faced with a global emerging disease threat, which highlighted its needs for improved international regulations governing the reporting of and response to unusual infectious-disease syndromes. Circulating air with high-efficiency particulate air (HEPA) filter to decontaminate, wearing masks and isolating a patient in a single room and wearing a gown, gloves, eye shield and mask or a portable air purifier which filters out small infectious particles (N95 mask) for staffs are necessary. Hui and Chan (2010) found that horseshoe bats are implicated in the emergence of novel coronavirus infection in humans. Ding et al., (2004) indicated that in addition to viral spread through a respiratory route, SARS-CoV in the intestinal tract, kidney and sweat glands maybe excreted via faces, urine and sweat, so leading to virus transmission. Wu and Yan (2004) consider the amino acid pairs as potential candidates for anti-SARS drugs, because they have greater chance of colliding with anti-SARS drugs, and they are more likely to link with the protein functions, also they are less vulnerable to mutations. The three-dimensional structure result indicates that the nsp2 protein of GD strain is high homologous with 3CL(pro) of SARS-CoV urbani strain, 3CL(pro) of transmissible gastroenteritis virus and 3CL(pro) of human coronavirus 229E strain, which further suggests that nsp2 protein of GD strain possesses the activity of 3CL(pro) (Lu et al., 2005). Rabenau et al., (2005) showed that SARS-CoV can be inactivated easily with commonly used disinfectants. Cao et al., (2016) showed that SARS transmission changes in its epidemiological characteristics and SARS outbreak distributions show palpable clusters on both spatial and temporal scales, also its transmission features are affected by spatial heterogeneity. Basic patient information and Spatial pattern of the SARS outbreaks are shown in Figure 3 and 4, respectively. Distribution of SARS cases during 2002-2003 in mainland China is shown in Figure 5. Electron micrograph of SARS-associated coronavirus is indicated in Figure 6.

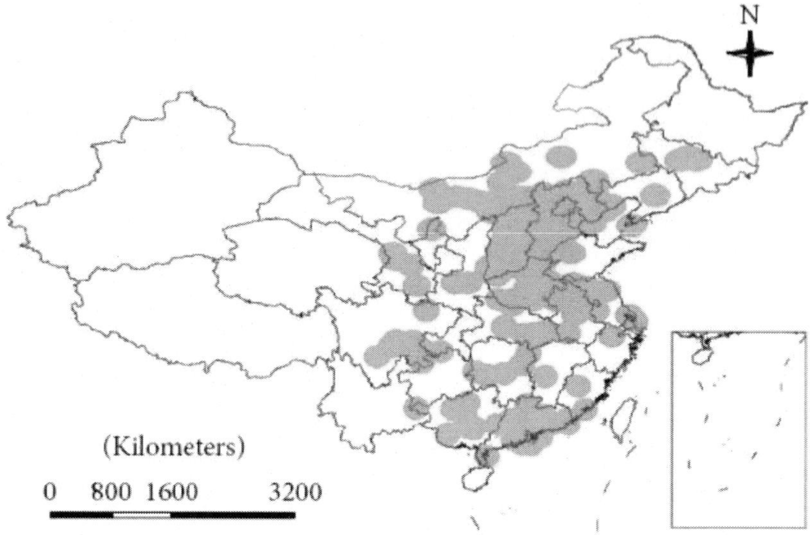

Figure 4. Spatial pattern of the SARS outbreaks (Cao et al., 2016).

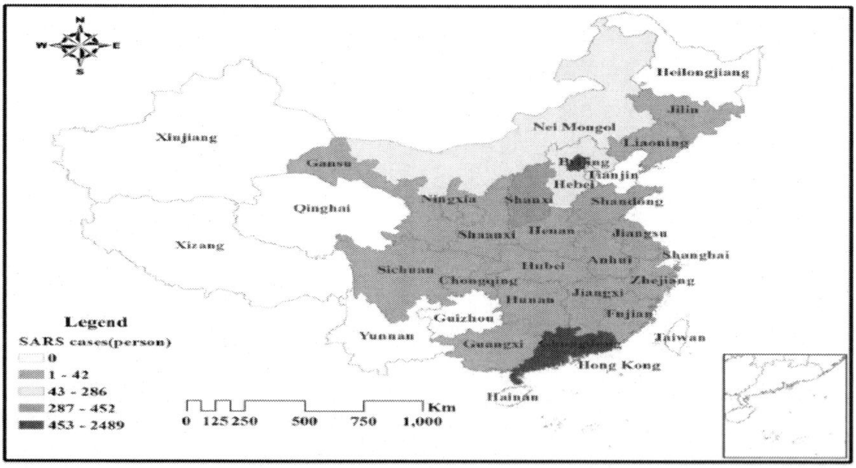

Figure 5. Distribution of SARS cases during 2002-2003 in mainland China (Xu et al., 2014).

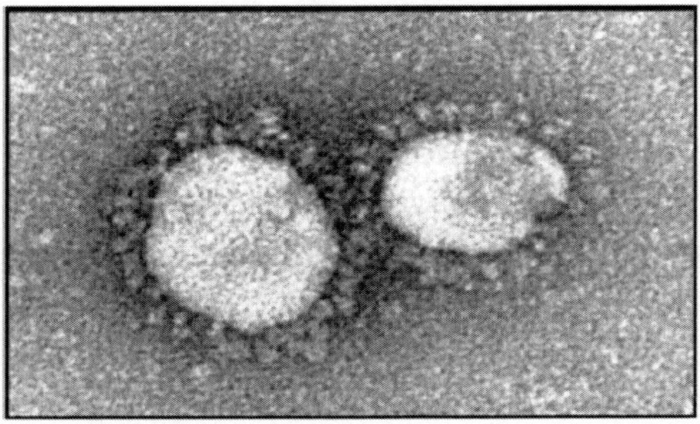

Figure 6. Electron micrograph of SARS-associated coronavirus (Parashar and Anderson, 2004).

Table 7. The most important treatments (Nie et al., 2003)

Ribavirin and gucocorticoid therapy
Antibacterial treatment (Levofloxacin and/or clarithromycin)
Ribavirin and methylprednisolone
Standard corticosteroid regimen for 21 days
Ribavirin regimen for 10-14 days
Pulsed methylprednisolone
Mechanical ventilation
Proteinase inhibitors
Intravenous immunoglobulins
Convalescent serum
Tumor necrosis factor alpha-blockers
Interferons
Traditional Chinese medicines

The most important primary measures are isolation, ribavirin, and corticosteroid therapy, mechanical ventilation, convalescent plasma, and others (Nie et al., 2003). Fuji et al., (2004) also recommend the usage of steroids and ribavirin for SARS treatment. Cheung et al., (2004) also confirmed that noninvasive positive pressure ventilation (NIPPV) is the most effective treatment of acute respiratory failure (ARF) in the patients with SARS studied. Netland et al., (2010) suggested that rSARS-CoV-ΔE is an effective vaccine candidate that might be useful if SARS recurred.

Wong et al., (2018) proposed that SARS-CoV may exploit the unique functions of proteins 8b and 8ab as novel mechanisms to overcome the effect of inefficient inducer of interferon (IFN) response during virus infection. Keyaerts et al., (2004) reported that chloroquine, an old antimalarial drug, may be useful for immediate use in the prevention and treatment of SARS-CoV infections.

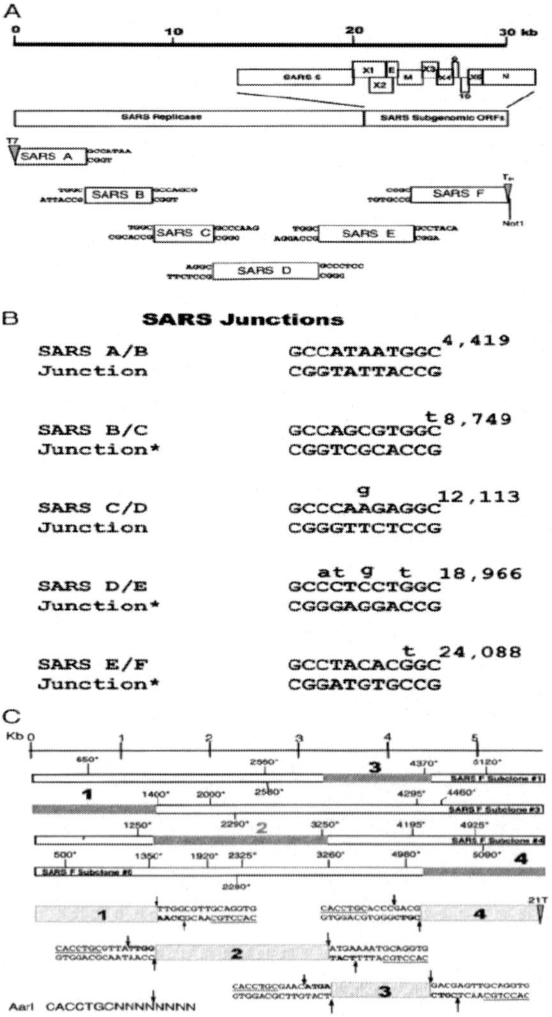

Figure 7. Assembly of a full-length SARS-CoV cDNA (Yount et al., 2003).

Table 8. Lessons learned from the SARS outbreak and concerns identified by WHO because of SARS

Lesson	Means	Concerns
The capacity of global alerts to improve awareness and viligance	Wide support by responsible press and amplified by electronic communications	Inadequate surge capacity in hospitals and public health systems
The advantage of quick detection and reporting	Immediate reporting of initial cases by South Africa and India	Healthcare providers themselves being the victims of the disease
The successful containment that can be achieved by readying health services with preparedness plans and campaigns to guard against imported cases	Climate of high alert that was established after reports of the disease became known	Shortage of expert staff to coordinate national and global responses to a rapidly evolving public health emergency
The value of immediate political commitment at the highest level	The experience in Vietnam, where the government took immediate measures to protect its people	In some cases, the need for hasty construction of new facilities; in other cases, hospitals being closed
The ability of even developing countries to triumph over a disease when reporting is prompt and open and when rapid case detection, immediate isolation and infection control, and vigorous contact tracing are put in place	The appeal by Vietnam, where WHO assistance was requested quickly and fully supported	The power of poorly understood infectious diseases to incite widespread public anxiety and fear, social unease, economic losses, and unwarranted discrimination

Demmler and Ligon, 2003; Maxwell et al., 2017.

Lau et al., (2009) discovered that the combination of ribavirin and cortico-steroids has no therapeutic benefits if it consumes early during SARS infection. The most important treatments for SARS is shown in Table 7. Lessons learned from the SARS outbreak and concerns identified by WHO because of SARS are presented in Table 8. Duration of clinical phases of the mild and moderately severe variants of severe acute respiratory syndrome are shown in Table 9. Assembly of a full-length

SARS-CoV cDNA is shown in Figure 7. Major mechanisms contributing to the pathogenesis of SARS are presented in Figure 8.

Table 9. Duration of clinical phases of the mild and moderately severe variants of severe acute respiratory syndrome

		Phase		
		Respiratory		
Time	Prodrome	Early	Late	Recovery
From onset, days	0	2-7	8-12	14-18
Duration, days	2-7	1-10	5-10	5-7

Christian et al., 2004.

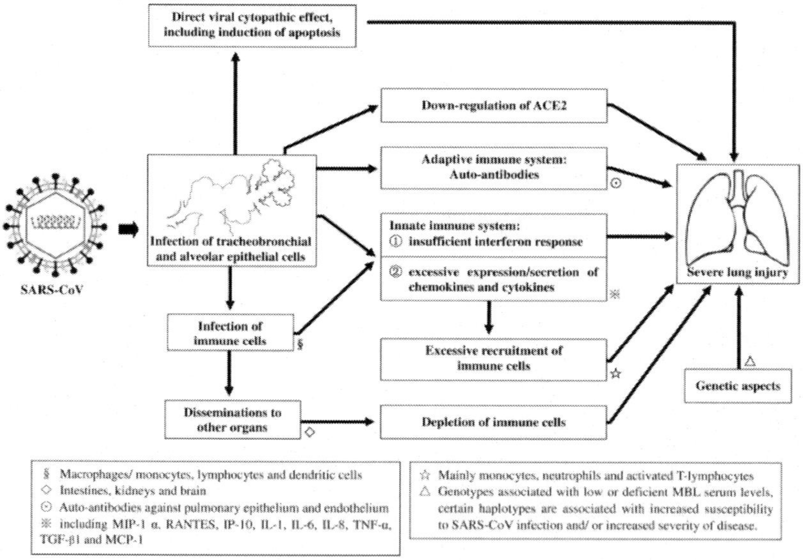

Figure 8. Major mechanisms contributing to the pathogenesis of SARS. These pathological events and cascade of changes from the basis of clinical symptoms and pathological findings at different stages of SARS. Correct recognizing of the pathogenesis may provide better guideline for prevention, diagnosis and treatment. MIP-1α, macrophage inflammatory protein-1α; RANTES, regulated on activation normal T cell expressed and sereted; TNF-α, tumor necrosis factor-α; TGF-β1, transforming growth factor-β1; MCP-1, monocyte chemoattractant protein-1 (Gu and Korteweg, 2007).

Chapter 8

THE MIDDLE EAST RESPIRATORY SYNDROME CORONAVIRUS (MERS-COV)

The middle East respiratory syndrome coronavirus (MERS-CoV) is a zoonotic beta coronavirus which can infect various kinds of animals such as humans, camels and bats (Hijawi et al., 2012; Chan et al., 2015, Al-Tawfiq and Memish, 2016; Kim et al., 2016; Jung et al., 2018; Ko et al., 2019). It belongs to the Betacoronavirus genus, of the coronavirus family (de Groot et al., 2013; Zaki et al., 2012). It was first discovered in September 2012 as the cause of death in a patients who had died of severe pneumonia in June 2012 in Jeddah, Saudi Arabia (Zaki et al., 2012; Eifan et al., 2017; Adegboye et al., 2019), and 1,348 cases of MERS-CoV infection confirmed globally, with at least 479 related deaths until June 23, 2015 (WHO, 2015). It has exported from the Middle East to other countries even countries in Asia, Europe and North America (WHO, 2015; Cha et al., 2018; Yavarian et al., 2018). Countries which are infected by MERS is shown in Table 1. Global map of confirmed MERS-CoV infections, 2012-2018 is presented in Figure 1. The most important characteristics of MERS-CoV infections is shown in Table 2.

Table 1. Countries which show cases

Most cases	Few Cases
Saudi Arabia, Oman, Kuwait, Qatar, United Arab Emirates, Jordan and Yemen (Al-Tawfiq and Memish, 2014; Almaghrabi and Omrani, 2017)	Iran, Bangladesh, Malaysia, Egypt, Tunisia, Algeria, the United Kingdom, France, Germany and the US (Mailles et al., 2013)

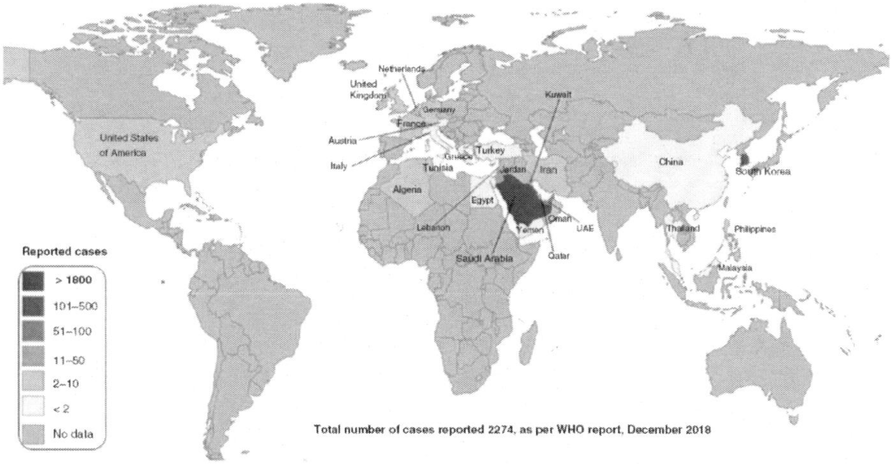

Figure 1. Global map of confirmed MERS-CoV infections, 2012-2018 (Choudhry et al., 2019).

Table 2. The most important characteristics of MERS-CoV infections

MERS-CoV infects more males than females (Assiri et al., 2013; Arabi et al., 2014)
1. It is accompanied by a cluster of flu-like symptoms (Al-Tawfiq et al., 2014; Noorwali et al., 2015; Shalhoub et al., 2015)
2. Life threatening severe illnesses consist of acute respiratory distress syndrome, pneumonia, myocarditis, and organ failure (Assiri et al., 2013; Arabi et al., 2014; Memish et al., 2013; Memish et al., 2014; Assiri et al., 2016)
3. Death occurs in 30% to 60% of the cases (Assiri et al., 2013)
4. Those who have pre-existing medical conditions such as diabetes, cardiovascular diseases, renal failure, obesity, and immunodeficiency are more vulnerable to this disease (Assiri et al., 2013; Al-Tawfiq et al., 2014)

Routine measures for travelers to help preventing the spread of viruses are hand washing, personal hygeien, avoid contacts with sick people, and animals, and cover the mouths with a tissue when coughing or sneezing and dispose properly the used tissue (Al-Tawfig et al., 2014). Because of inadequate implementation of a quarantine protocol for close contacts and poor public health surveillance, one case was exported to Huizhou, China via Hong Kong during the early outbreak (Hui et al., 2015). MERS-CoV is most likely derived from an ancestral reservoir bats (van Boheemen et al., 2012; Cui et al., 2013; de Wit and Munster, 2013; Corman et al., 2014; Munster et al., 2016). Except for some cases in Korea in 2015, 82% of infections have occurred in Saudi Arabia, and the human mortality rate of MERS-CoV infection was nearly 35% (Meyerholz et al., 2016; WHO, 2016; Alagaili et al., 2019). It has been reported that patients with severe diseases have at least one underlying condition, including diabetes, hypertension, chronic cardiac disease and chronic renal disease (Assiri et al., 2013; Poissy et al., 2014; Faridi, 2018). Human to human transmission has been facilitated in healthcare settings (Assiri et al., 2013; Amer et al., 2018; Park et al., 2019) with the contribution of hospital-based transmission of MERS estimated at about 80% using an epidemic model (Chowell et al., 2014). MERS outbreak was found in the Republic of Korea since 2015 (Cowling et al., 2015; Hui et al., 2015), which showed the importance risk of importing MERS and escalate global spread of MERS and damage to both economic and public health activities (Kucharski and Althaus, 2015; Al-Jasser et al., 2019), which show the importance of travel restrictions for infected countries (Poletto et al., 2014; Bogoch et al., 2015). The key receptor for MERS-CoV infection which is dipeptidyl peptidase 4(DPP4), is widely distributed on human endothelial and epithelial cells (Meyerholz et al., 2016). Coronavirus entry is initiated by the binding of the spike protein (S) to cell receptors, specifically, dipeptidyl peptidase 4 (DDP4) and angiotensin converting enzyme 2 (ACE2) for MERS-CoV and SARS-CoV, respectively (Coleman and Frieman, 2014; Greenberg, 2016; De Wit et al., 2016). Its genome is single-stranded RNA which encodes 10 proteins including two replicase polyproteins (open reading frames [ORF], 1 ab and 1 a), three structural proteins (E, N, and M), a surface glycoprotein (S, spike) which comprises S1 and S1, and five nonstructural proteins (ORF 3, 4a, 4b, and 5) (Wang et al., 2013; Zumla et al., 2016; Wernery et al., 2017). It has been concluded that the MERS-S protein is known to represent a key target for the development of new therapeutics and includes of a receptor-binding subunit S1 and a membrane-fusion subunit S2 (Du et al., 2017). The subunit S1 is composed of four different core domain (Haverkamp et al., 2019), and the domain S1B binds to the host-cell receptor dipeptidylpeptidase 4 (DPP4) (Lu et al., 2013; Raj et al., 2013; Wang et al., 2013; Inn et al., 2018), while the domain S1A binds to sialoglycans which increased infection of human lung cells by MERS-CoV (Li et al., 2017). The roles of S protein in receptor binding and membrane fusion make it a perfect target for vaccine and antiviral development (Wang et al., 2016). It has been showed that vaccines based on the S protein can induce antibodies to block virus binding and fusion or neutralize virus infection (Du et al., 2009; Zhang et al., 2014; Papaneri et al., 2015). The subunit S2 includes the fusion peptide, two heptad repeats and a transmembrane domain, which mediate fusion of the virus with the cell membrane (Lu et al., 2013; Raj et al., 2013; Wang et al., 2013; Pallesen et al., 2017). Antibodies against MERS-CoV have been found among both dromedary camel populations and camel-exposed humans, which have been recognized as the source of multiple MERS case importations around the world (Meyer et al., 2014; Muller et al., 2015; Haverkamp et al., 2019). Previous studies have investigated viral vector-based vaccines (Song et al., 2013; Kim et al., 2014; Guo et al., 2015; Volz et al., 2015;Alharbi et al., 2017), subunit vaccines (Du et al., 2013; Ma et al., 2014; Tang et al., 2015; Zhang et al., 2015; Coleman et al., 2017), and DNA vaccines (Muthumani et al., 2015, Wang et al., 2015), but there is no clinically approved vaccine for MERS-CoV (Jung et al., 2018). Among these vaccines, viral vectors or DNA immunization successfully generated neutralizing antibodies and protected against infection (Alharbi et al., 2017). Na and Dabir (2006) reported that subunit vaccines could not induce enough immune responses in host, resulting in failure to make long-term memory of antigen.

Table 3. Timing of documented case importations of Middle East respiratory syndrome (MERS) around the world

Country	Date of arrival	Days since 3 September 2012
United Kingdom	2012/9/11	8
Germany	2012/10/24	51
United Arab Emirates	2013/3/8	186
France	2013/4/17	226
Tunisia	2013/5/3	242
Italy	2013/5/25	264
Oman	2013/10/26	418
Kuwait	2013/11/7	430
Yemen	2014/3/17	560
Malaysia	2014/4/7	581
Philippines	2014/4/15	589
Greece	2014/4/17	591
Jordan	2014/4/19	593
Lebanon	2014/4/22	596
United States	2014/4/24	598
Egypt	2014/4/25	599
Iran	2014/5/1	605
Netherlands	2014/5/10	614
Algeria	2014/5/28	632
Austria	2014/9/22	749
Turkey	2014/10/6	763
South Korea	2015/5/4	973
China	2015/5/26	995
Thailand	2015/6/15	1015

Nah et al., 2016

Jung et al., (2018) found that heterologous prime-boost may induce longer-lasting immune responses against MERS-CoV because of an appropriate balance of Th1/Th2 responses, and both heterologous prime-boost and homologous spike protein nanoparticles vaccinations could provide protection from MERS-CoV challenge in mice. They finally demonstrated that heterologous immunization by priming with Ad5/MERS and boosting with spike protein nanoparticles could be an efficienct prophylactic strategy against MERS-CoV infection. Early detection and isolation of patients with MERS-CoV infection remains an important parameter for the control of MERS-CoV transmission (Memish and Al-Tawfiq, 2014; Al-Tawfiq and Perl, 2015; Ahmadzadeh et al., 2020). Timing of documented case importations of Middle East respiratory syndrome (MERS) around the world is presented in Table 3. The most important symptoms of MERS from the highest to the lowest are presented in Table 4. Countries at high risk of case importations of Middle East respiratory syndrome (MERS) are shown in Figure 2.

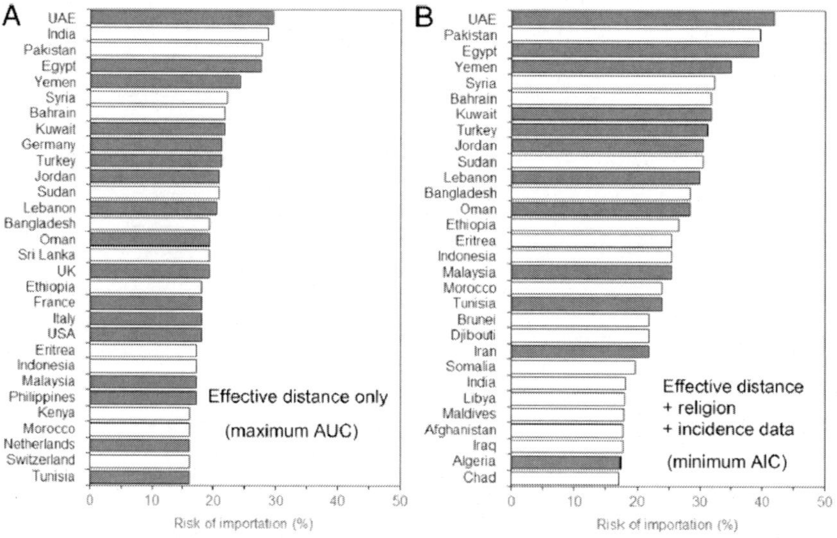

Figure 2. Countries at high risk of case importations of Middle East respiratory syndrome (MERS). List of 30 countries with the estimated highest importation risks by 3 September 2015. The panel a shows the prediction that used the effective distance only with the best predictive value as assessed by the area under the curve (AUC; model 1). The panel b shows the prediction using a model that used the effective distance as well as religion and incidence data of MERS in the Saudi Arabia (Nah et al., 2016).

Four coronavirus genera have been identified including alpha- (group 1), beta- (group 2), gamma- (group 3), and deltacoronavirus (group 4). HCoVs are among the alphacoronavirus and betacoronavirus (Wang et al., 2018). Phylogenetic analysis of coronaviruses based on complete genomes is shown in Figure 3. Genomic Mapping of MERS-CoV is indicated in Figure 4.

Figure 3. Phylogenetic analysis of coronaviruses based on complete genomes (Wang et al., 2018).

Table 4. The most important symptoms of MERS from the highest to the lowest

Fever
Cough
Shortness of breath
Sore throat
Diarrhea
Head and body aches
Vomiting
Chest pain/tightness
Running nose
Altered conscious/confusion
Sweating
Abdominal pain
Weakness/fatigue
Dizziness
Loss of appetite
Shivering

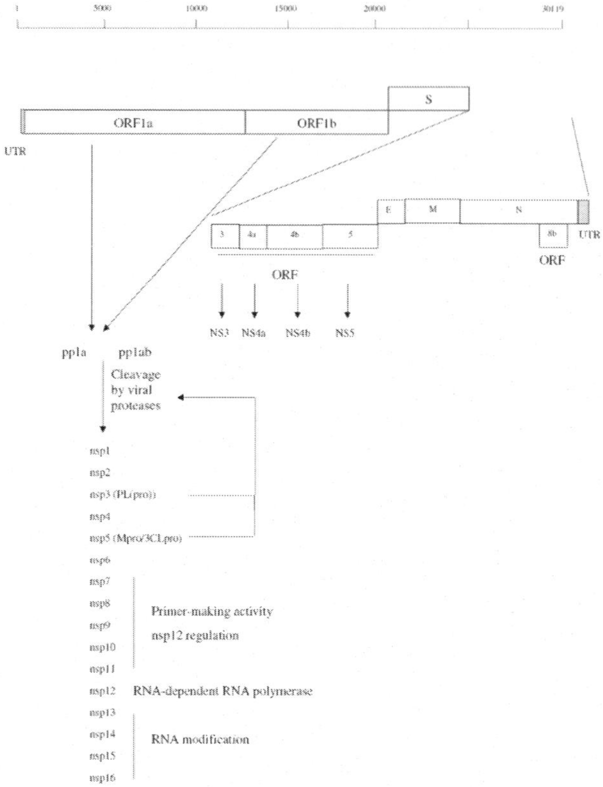

Figure 4. Genomic Mapping of MERS-CoV (Al-Omari et al., 2019).

The 5′ end contains the rep1a and rep1b genes, which encode the viral replicase-transcriptase. At the 3′ end of genome, four structural protein, namely spike (S), envelope (E), membrane (M), and nucleocapsid (N) protein and five accessory proteins (ORF3, ORF4a, ORF4b, ORF5 and ORF8) make up 10 kb. M and E proteins play important role in viral assembly, N protein is required for RNA synthesis (Wang et al., 2018; Masters, 2019). The MERS-CoV genome structure and virion is shown in Figure 5. Kasem et al. (2018) and Dighe et al. (2019) showed that camels are a main reservoir for the maintenance of MERS-CoVs, and they are an important source of human infection with MERS. Phylogenetic analysis of MERS-CoV found in ten RT-PCR-positive camels is presented in Figure 6.

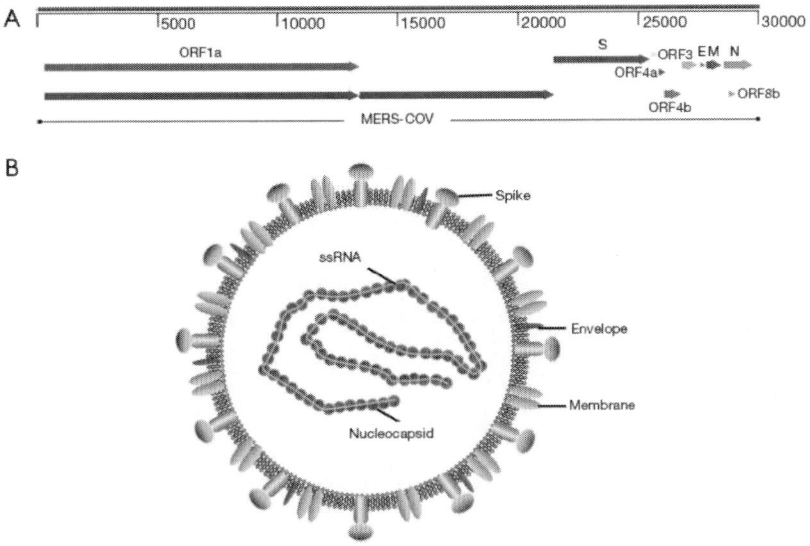

Figure 5. The MERS-CoV genome structure and virion (Wang et al., 2018).

Al-Tawfiq et al. (2017) found that MERS-CoV was a rare cause of community acquired pneumonia and other viral caused such as influenza which were more common. Alfaraj et al. (2019) found that different factors contributed to increase mortality rate for MERS-CoV patients, and one of the most important factor is usage of corticosteroid and a continuous renal replacement therapy (CRRT). Corman et al. (2014) found that the kit is important tool for assisting in the rapid diagnosis, patient management and epidemiology of suspected MERS-CoV cases. Yoon et al. (2019) found that 6,8-difluoro-3-isobutyryl-2-((2,3,4-trifluorophenyl)amino) quinolin-4(1H)-one (6u) shows high inhibitory influence and low toxicity activities which is from 3-Acyl-2-phenylamino-1,4-dihydroquinolin-4(1H)-one derivatives. Qiu et al. (2016) indicated that hMS-1 might be developed as an effective immunotherapeutic agent to cure patients infected with MERS-CoV, especially in emergent cases. Early MERS-CoV diagnosis may require more sensitive risk assessment tools to reduce avoidable delays, specifically those related to patients and health system (Park et al., 2015; Ahmed, 2017; Lee and Cho, 2017; Ahmed, 2019; Tang et al., 2020).

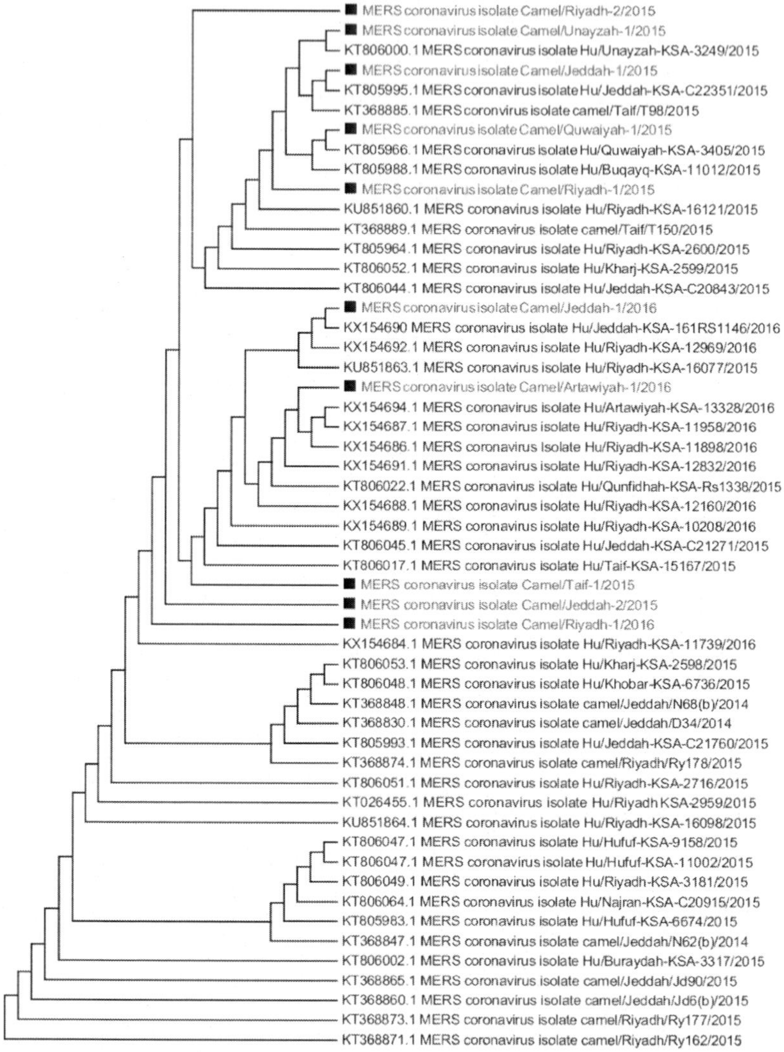

Figure 6. Phylogenetic analysis of MERS-CoV found in ten RT-PCR-positive camels, using MEGA7. Full genome sequences of the ten MERS-CoV camel samples and the sequences from corresponding patients were aligned with sequences of MERS-CoV reference strains available from GenBank. Phylogenetic analysis was inferred using the neighbor-joining method and distance calculations were computed using the Tamura-Nei model. Sequences from the current study are indicated by a soil square (Kasem et al., 2018).

Badawi and Ryoo (2016) showed that protection against MERS-CoV and other respiratory infections can be increased if public health vaccination strategies are tailored to target persons with chronic disorders. Alqahtani et al., (2017) noted that still many people have lack of accurate understanding about MERS-CoV transmission and prevention, and more studies need to examine the knowledge and practices among public and workers about MERS-CoV (Khan et al., 2014; Almutairi et al., 2015; Kharma et al., 2015; Al-Mohrej et al., 2016; Bawazir et al., 2018; Lee et al., 2018; Alfaraj et al., 2019). Close contacts include airplane setting, household setting, household setting who also visited the patient in hospital, healthcare setting (Al-Tawfiq, 2013; Hashem et al., 2019). Al-Tawfiq and Auwaerter (2019) suggested proper infection control procedures, prompt recognition, isolation and management of suspected cases are important parameters for MERS prevention. Kim (2018) mentioned the importance of protecting healthcare providers from severe both physical and psychological stress. Al-Tawfiq et al., (2019) reported notable increase in costs of the healthcare system because of increase in utilization of surgical masks, respirators, soap, and alcohol-based hand sanitizers Douglas et al., (2018) indicated that the extent of MERS-CoV adaptation determines the minimal infectious dose needed to achieve severe respiratory disease. Nikiforuk et al., (2016) reported that viral infectious clone system may shorten time between emergence of a novel viral pathogen and construction of an infectious clone system. Ebihara et al., (2005) found that virus infectious clone systems allow for expression of a homogenous virus population within mammalian cell culture from a sequence of DNA or RNA. Letko et al., (2018) demonstrated that MERS-CoV spike can utilize multiple paths to rapidly adapt to novel species variation in DPP4. Zhang et al., (2018) found that MERS-4 neutralizes MERS-CoV by indirect rather than direct competition with DPP4. Coleman et al., (2017) concluded that the MERS-CoV nanoparticle vaccine produced high titer anti-S neutralizing antibody and protected mice from MERS-CoV infection *in vivo*. Mustafa et al., (2018) suggested application of antimicrobial peptides (AMPs) as alternative therapeutic agents against MERS-CoV infection.

Table 5. Comparison between SARS and MERS-CoV in respect to their virology, epidemiology and clinical outcomes

	MERS-CoV	SARS
Virology	Betacoronavirus lineage 2C	Betacoronavirus lineage 2B
Receptor	hDPP4	ACE2
Genome size	29.9 kb	29.3kb
Source	Not yet confirmed, camel is the likely host	Civet Cat
Epidemiology	Limited human to human transmission, the disease is mostly localized in the Middle East	Human to human transmission is well-recognized, affected many countries but spared the Middle East
Ro	2-3(for Jeddah 3.5-6.7, for Riyadh 2-2.8)	Variable, ranges from 2-6
Superspreading event	Not known	Reported
M:F	1.74:1	0.75:1
Median age (range) in years	48 (1-99)	Less than a third had
Mean incubation period in days (range)	5 (2-15)	Comorbidities
Comorbidities	Three quarter of the patients had comorbidities	Less than a third has Comorbidities
Clinical presentation	Unpredictable and erratic clinical course ranging from asymptomatic illness to severe pneumonia	A typical biphasic clinical course
Haemoptysis	More common	Less common
Respiratory failure	Presents relatively early	Presents relatively late
Travel association	Limited travel-associated exposure	Recognized travel-associated exposure
Time from symptom onset to hospitalization	0-16 days	2-8 days
Median time from symptom onset to death	12 days	21 days

Banik et al., 2015.

Table 6. The difference between MERS-CoV and SARS-CoV in symptoms, signs, laboratory tests, and chest film

Symptoms	**MERS-CoV**	**SARS-CoV**
Headache	+	++
Fever and chills	+++	++
Prominent fatigue	+	-
Myalgias	++	+++
Dry cough	+++	++
Shortness of breath	+++	++
Sore throat	+	+
Nausea/vomiting	+	+
Diarrhea	±	±
Abdominal pain	±	-
Hemoptysis	±	-
Signs		
Tachycardia	+	+
Conjunctival suffusion	+	-
Diminished breath sounds	+	+
Acute renal failure (ARF)	±	-
Laboratory tests		
Normal WBC count	-	
Leukopenia	+	-
Relative lymphopenia	+++	+++
Thrombocytopenia	++	+++
Elevated serum transaminases	+	±
Elevated ldh	+	++
Elevated cpk	++	-
Chest film		
Normal/minimal basilar infiltrates (early)	-	+
Unilateral infiltrates (late)	+	-
Pleural effusion	+++	++
Cavitation	+	-
ARDS (severe cases)	-	-

Cunha and Opal, 2014.

Table 7. Enhanced infection control measures that were effective in controlling nosocomial outbreaks

1. Hand hygiene, and droplet and contact precautions for febrile patients with a fever before testing these patients for MERS-CoV.
2. Putting surgical masks on all patients undergoing haemodialysis, and ensuring health-care workers wear N95 filtering facepiece respirators when managing any patient with a confirmed MERS-CoV infection who is undergoing an aerosol-generating procedure.
3. Patients with suspected MERS-CoV infection admitted to dialysis or intensive care units should be placed in isolation rooms with a portable dialysis machine.
4. Increasing environmental cleaning, and preventing non-essential staff and visitors from coming into contact with patients infected with MERS-CoV.

Hui et al., 2018.

Table 8. Potential therapies for the treatment of MERS coronavirus infection

Antibody-based interventions	Whole blood
	Convalescent plasma
	Intravenous immunoglobulin
	Polyclonal human immunoglobulin (SAB-301) from transgenic cows
	Equine antibody fragments
	Camel antibodies
	Monoclonal antibodies (e.g., MERS-4, MERS-7)
	Human monoclonal antibodies to Sprotein
	Humanised anti-S monoclonals (e.g., hMS-1, m336, 4C2)
Interferons	Interferon alfa (1a,2b)
	Interfron beta-1b
	Interferon gamma
Antivirals	Ribavirin monotherapy (with or without interferon)
	HIV protease inhibitors (e.g., lopinavir, nelfinavir)
	Cyclophilin inhibitors (e.g., ciclosporin, alisporivir)
	Nucleoside viral RNA polymerase inhibitors (galidesivir, remdesivir, EIDD 2801)
MERS-CoV 3C-like protease inhibitors	Flavinoids (e.g., herbacetin, isobavachalcone, quercetin 3-β-d-glucoside, helichrysetin)
Combination therapies	Lopinavir-ritonavir and interferon beta 1b
	Cyclosporin plus interferon alpha
	Lopinavir-ritonavir
	Ribavirin and interferon alpha

Host-direct therapies (repurposed drugs)	Chlorpromazine
	Mycophenolic acid (with or without interferon 1b)
	Nitazoxanide
	Sitagliptin
	Omacetaxine mepesuccinate
	Aciclovir
	Imatinib mesylate (tyrosine mesylate inhibitor)
	Neurotransmitter inhibitors (e.g., clomipramine, astemizole)
	Neutriceuticals (e.g., zinc)
Host-directed therapies (cellular therapy)	Allogeneic mesenchymal stromal cells
Lectins	Mannose binding lectin
Anti-coronavirus peptides	Peptides derived from HR1, HR2, and RBD spike protein subunits
	Peptides inhibiting viral entry and replication
	Peptides derived from antimicrobial peptides
RNA interference molecules silencing key MERS-CoV genes	miRNA molecules
	siRNA molecules

MERS= Middle East respiratory syndrome, MERS-CoV= Middle East respiratory syndrome coronavirus, RBD= Receptor-binding domain, S= Spike, RNA= Ribonucleic acid, miRNA= micro RNA, siRNA= small interfering RNA.

Arabi et al., 2015; Zumla et al., 2016; Rabaan et al., 2017; Sheahan et al., 2017; Arabi et al., 2018; Mustafa et al., 2018; Agostini et al., 2019; Behzadi and Leyva-Grado, 2019; Beigel et al., 2019; Kim et al., 2019; Momattin et al., 2019; Zhou et al., 2019; Memish et al., 2020.

Baharoon and Memish (2019) emphasized on balance in application of both vaccination and antiviral therapeutics, also they highlighted the importance of avoid mechanism of escape mutant virus strains and improve activity against divergent virus strains. Widagdo et al., (2017) reported vaccination of dromedary camels is an appropriate way to drop in human MERS cases. Lu et al., (2015) indicated that both SARS and MERS contain a surface-located spike (S) protein which initiates infection by mediating receptor-recognition and membrane fusion which is a key factor in host specificity. Liu et al., (2020) found that pregnant women are more susceptible to respiratory pathogens, due to the characteristic immune responses during pregnancy and potential risks from the cytokine-storm by COVID-19 infections.

Table 9. Middle East respiratory syndrome coronavirus vaccines

Vaccine	Target	Use	Advantages	Problems
Anti-MERS-CoV monoclonal antibodies	Surface (S) glycoprotein	Passive immunization; prophylaxis or treatment at early times p.i.	High titer preparations; can be produced in large amounts	Short half-life; needs to be re-administered for continued efficacy
Human polyclonal anti-MERS-CoV antibodies	Virus structural proteins	Passive immunization; treatment at early times p.i.	Polyclonal antibody so antibody escape unlikely; human antibody	Short half-life; needs to be re-administered for continued efficacy; few MERS survivors available as donors
Inactivated virion vaccines	Virus structural proteins; anti-S neutralizing antibodies most important	Active immunization	High titer antibody to S protein	Response many not be long term; on challenge may induce immunopathological disease; may be ineffective in aged populations
Live attenuated vaccines (e.g., viruses deleted in envelope (E) protein; viruses with reduced fidelity (mutated in nsp14)	Mostly virus structural proteins	Active immunization	Generally safe; induce antibody and T-cell responses; long-term immunity	May not be safe is immunocompromised patients; may regain virulence by reversion or recombination with circulating CoV
Viral vector (attenuated) vaccines: poxvirus, AAV adenovirus, parainfluenza virus, rabies virus, measles virus, VSV	S protein	Active immunization	Safe: non-replicating; induce antibody and T-cell responses	Long-term immunity, but not as long as live attenuated vaccines

Vaccine	Target	Use	Advantages	Problems
Replicon particles (e.g., VEEV or VSV-based)	S protein or any viral protein	Active immunization	Safe; non-replicating; induce antibody and T-cell responses; useful for mucosal immunity	Production is complex
Subunit vaccines (e.g., RBD of S protein)	Generally S protein	Active immunization	Safe; non-replicating; induce high antibody titers; may also induce T-cell responses	Duration of response not known
DNA vaccines	Generally S protein	Active immunization	Safe; induce high antibody titers and T-cell responses	Immunogenicity variable; may induce anti-DNA immune response

MERS-CoV, Middle East respiratory syndrome coronavirus; p.i., post infection; AAV, adeno-associated virus; VSV, vesicular stomatitis virus; VEEV, Venezuelan equine encephalitis virus; RBD, receptor binding domain.

Perlman and Vijay, 2016

Table 10. Drug regimes used in the treatment of SARS and MERS

Treatment plant for SARS	MERS
Ribavirin (Oral/IV) Antibiotics ± corticosteroids ± immunoglobulin	Ribavirin (Oral/IV) IFN-α2b Corticosteroids
Ribavirin (Oral/IV) Lopinavir/Ritonavir ± corticosteroids	Ribavirin (Oral/IV) PEGylated IFN-α2a (IV) ± corticosteroids
IFN-alfacon-1 ± corticosteroids ± antibiotics	Ribavirin (Oral/IV) Lopinavir/Ritonavir IFN-α2b
Fluoroquinolone (IV) Azithromycin (IV) IFN-α (IM) ± corticosteroids ± Immunoglobulins ± thymic peptides/proteins	
Quinolone (IV) Azithromycin (IV) ± IFN-α ± corticosteroids	
Levofloxacin Azithromycin ± IFN-α ± corticosteroids	

Abbreviations: IFN interferon, IM intramuscular, IV intravenous, SARS severe acute respiratory syndrome.

Dyall et al., 2017.

Comparison between SARS and MERS-CoV in respct to their virology, epidemiology and clinical outcomes are shown in Table 5. The difference between MERS-CoV and SARS-CoV in symptoms, signs, laboratory tests, and chest film is indicated in Table 6. Enhanced infection control measures that were effective in controlling nosocomial outbreaks is shown in Table 7. Potential therapies for the treatment of MERS coronavirus infection are shown in Table 8. Middle East respiratory syndrome coronavirus vaccines are shown in Table 9. Drug regimes used in the treatment of SARS and MERS are shown in Table 10. Adverse effects of Ribairin are presented in Table 11.

Table 11. Adverse effects of Ribavirin

1. Haemolytic anaemia, leukopaenia, thrombocytopaenia
2. Reticulocytosis
3. Hypocalcaemia, hypomagnesaemia, hypokalaemia, hyperuricaemia
4. Bradycardi
5. Nausea, vomiting, diarrhoea
6. Chest pain, dyspnoea, rhinitis, pharyngitis
7. Sleep disturbance, irritability, arthhralgia, paraesthesia, blurring of vision
8. Influenza-like illness, dizziness, arthralgia, headache
9. Thyroid disease
10.Rash, pruritis, alopecia, dry skin
11.Taste disturbance
12.Teratogenicity

Tai, 2007.

Chapter 9

COVID-19 CORONAVIRUS

The novel SARS-CoV-2 coronavirus which was started in the city of Wuhan, China, has caused a large scale COVID-19 epidemic and spread in all over the world (Bal et al., 2020). It is called SARS-CoV-2 because of its high similarity in terms of clinical symptoms and biological nature with the causative agent of severe acute respiratory syndrome (SARS) by the International Committee on Taxonomy of Viruses (Abduljali and Abduljali, 2020; Benvenuto et al., 2020), which can affect patients of all ages (Hsih et al., 2020; Lai et al., 2020). Its genome sequence analysis has shown that SARS-CoV-2 belons to betacoronavirus genus, which includes Bat SARS-like coronavirus, SARS-CoV and MERS-CoV (Petrosillo et al., 2020). Its outbreak in China since December 2019 has caused so many challenges, and it has rapidly spread to many countries (Robson, 2020; Roda et al., 2020; Wang et al., 2020). On March 11, the World Health Organization (WHO) declared a new pneumonia outbreak a global pandemic (Li et al., 2020). It belongs to a large family of viruses which are known as coronaviruses (Chen et al., 2020). It has emerged after 2003 Severe Acute Respiratory Syndrome (SARS) in China and the second coronavirus which was famous as the Middle East Respratory Syndrome (MERS) in 2012 in Saudi Arabia. On the basis of nucleic acid sequence similarity, the newly identified 2019-nCoV is a betacoronavirus (Chen et al., 2020; Yu Jun et al., 2020). It is mainly associated with respiratory disease and few extrapulmonary signs (Gralinski and Menachery, 2020; Lupia et al., 2020). Epidemiological investigations showed that different animals (Bats, pangolins, snakes) could have been intermediate hosts which facilitate the spill-over of SARS-CoV-2 as a distinct human Betacoronavirus from bats to human population (Cyranoski, 2020; Ji et al., 2020; lippi and Plebani, 2020; Lu et al., 2020; Yuan et al., 2020). Sun et al. (2020) reported that s bats are consider as the natural hosts of this virus, cold temperature and low humidity provides conducive environmental conditions for prolonged viral survival in suspected regions concentrated with bats. The resulting genomic sequence data has shown that the epidemic and the number of COIV-19 cases have been increasing due to human to human transmission just after a single introduction into the human population. The RBD portion of the SARS-CoV-2 pike proteins has evolved to effectively target a molecular feature on the outside of human cells called ACE2, a receptor involved in regulating blood pressure. The SARS-CoV-2 spike protein was found so effective at binding the human cells. The SARS-CoV-2 backbone differed substantially from those of known coronaviruses and mostly resembled related viruses found in bats and pangolins. With considering tabular comparison of SARS versus COVID-19, clinical presentation of SARS and COVID-19 are fever, dry cough, and shortness of breath; incubation period for SARS and COVID-19 are 2-7 days, 2-14 days, respectively (Sohrabi et al., 2020). Its cycle starts when S protein binds to the cellular receptor ACE2; after receptor binding the conformation change in the S protein facilitates viral envelope fusion with the cell membrane through the endosomal pathway, and then SARS-CoV-2 releases RNA into the host cell. Genome RNA is translated into viral replicase polyproteins pp1a and 1ab, which are then cleaved into small products by viral proteinases. The polymerase produces a series of subgenomic mRNAs by discontinuous transcription and finally translated into relevant viral proteins. Both viral proteins and genome RNA are subsequently assembled into virions in the endoplasmic reticulum (ER) and Golgi, and then transported via vesicles and released out or the cell

(Shereen et al., 2020). Kang et al. (2020) reported that the cause and consequence of pneumonia sill remain unknown. Previous coronavirus outbreaks have started, with humans contracting the virus after direct exposure to civets (SARS), and camels (MERS) (Andersen et al., 2020).

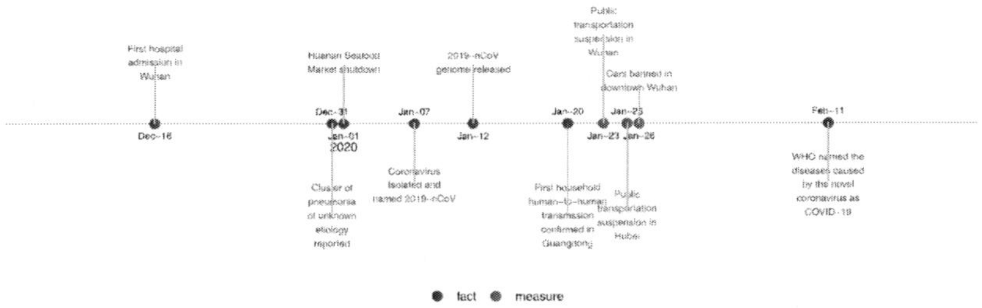

Figure 1. The timelines of the facts of COVID-19 and control measures implemented in Wuhan, China from December 2019 to February 2020. The red dots are the events in the COVID-19 outbreak, and the blue dots are the control measures (Lin et al., 2020).

Tseng et al. (2012) found that SARS-CoV vaccines all induced antibody and protection against infection with SARS-CoV. Most CoVs share a similar viral structure, similar infection pathway and a similar structure of the S protein (Yuan et al., 2017; Cheng et al., 2020), which has shown similar research strategies should be used for 2019-nCoV (Coutard et al., 2020; Yu et al., 2020). Matsuyama et al., (2020) discovered that SARS and MERS and SARS-CoV-2 infection is enhanced by TMPRSS2. Rasmussen et al., (2020) reported an incubation period of ~5 days (range-2-14 days), average age of hospitalized patients has been 49-56 years, with a third to half with an underlying illness, and men were more frequent among hospitalized cases (54-73%). Chan et al., (2020) reported person-to-person transmission of 2019 novel coronavirus in hospital and family settings, and the reports of infected travelers in other geographical regions. Khan et al., (2020) reported that during control the 2019n-CoV, the entrances of residential communities, dormitories and public places were restricted and temperature monitoring for related symptoms was done before residents entering. The timelines of the facts of COVID-19 and control measures implemented in Wuhan, China from December 2019 to

February 2020 is shown in Figure 1. The systemic and respiratory disorders caused by COVID-19 infection is presented in Figure 2. Genomic organization and virus structure of 2019-nCoV are presented in Figure 3.

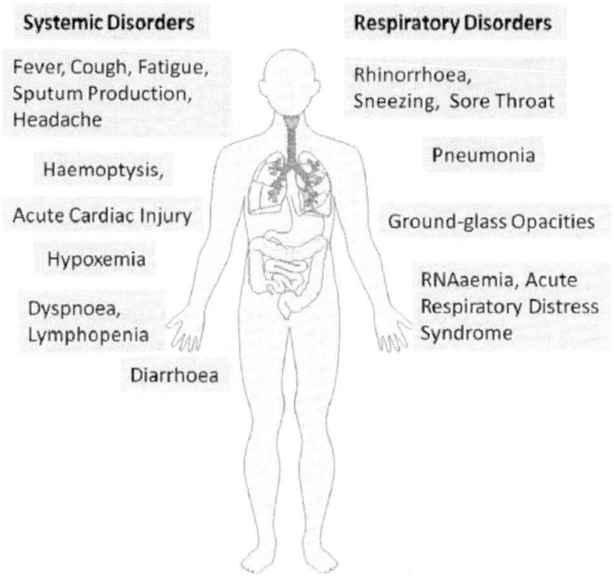

Figure 2. The systemic and respiratory disorders caused by COVID-19 infection. There are general similarities in the symptoms between COVID-19 and previous diseases, however, unlike COVID-19, only a low percentage of MERS-CoV or SARS-CoV patients exhibited diarrhoea (Rothan and Byrareddy, 2020).

Chen et al. (2020) showed that on the basis of molecular modeling, 2019-nCoV RBD has a stronger interaction with angiotensin converting enzyme 2 (ACE2), and a unique phenylalanine F486 in the flexible loop plays an important role due to its penetration into a deep hydrophobic pocket in ACE2, and ACE2 can potentially bind RBD of 2019-nCoV, making them all possible natural hosts for the virus. It might be able to bind to the angiotensin-converting enzyme2 receptor in humans (Lu et al., 2020). Luan et al. (2020) observed that N82 in ACE2 indicated a closer contact with SARS-CoV-2 S protein than M82 in human ACE2. RBD domain of SARS-CoV-2 interacts with human ACE2, which is why ACE2 is considered as the receptor for SARS-CoV-2 (Luan et al., 2020; Wall et al., 2020).Wall et al. (2020) showed the presence of a four amino acid residue insertion at the boundary between the S1 and S2 subunits in SARS-CoV-2 S compared with SARS-CoV and SARSr-CoV S. Wall et al. (2020) demonstrated that SARS-CoV s murine polyclonal antibodies potently inhibited SARS-CoV-2 S mediated entry into cells which showed cross-neutralizing antibodies targeting conserved S epitopes can be elicited upon vaccination. Tian et al. (2020) reported the measures to prevent transmission is very successful at early stage, and in the next steps on the COVID-19 infection should be focused on early isolation of patients and quarantine. Comparison of the known structures SARS-CoV RBD and its ACE2 complex with the deduced molecular model of 2019-nCoV RBD is shown in Figure 4. The emergence of SARS-CoV-2 and the outbreak of COVID-19 are presented in Figure 5. Possible origins of SARS-CoV-2 are shown in Table 1.

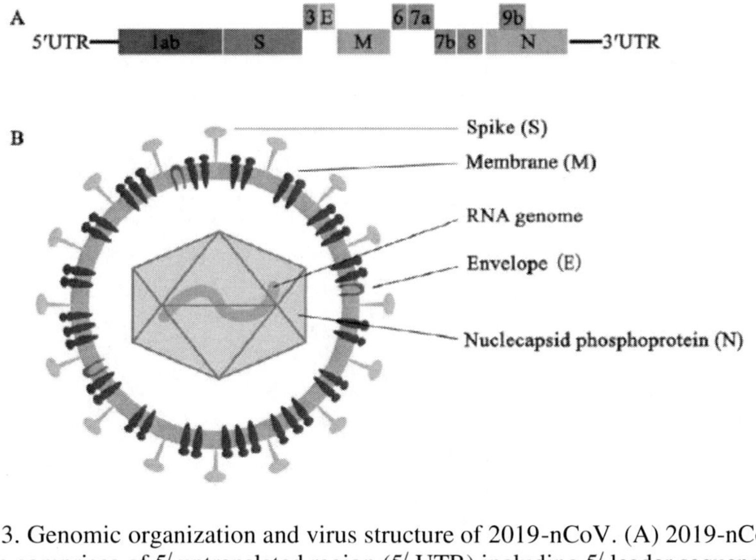

Figure 3. Genomic organization and virus structure of 2019-nCoV. (A) 2019-nCoV genome comprises of 5′ untranslated region (5′ UTR) including 5′ leader sequence, open reading frame (ORF) 1a/b, envelop, membrane and nucleoprotein, accessory proteins such as orf 3, 6, 7a, 7b, 8 and 9b and 3′ untranslated region (3′ UTR) in sequence. (B) 2019-nCoV structure consists of single-strand, positive-sense RNA as the genetic material surrounded by nucleocapsid protein in the core and an envelope containing four proteins: Spike protein, Envelope protein, Membrane protein (Han et al., 2020; Sigrist et al., 2020).

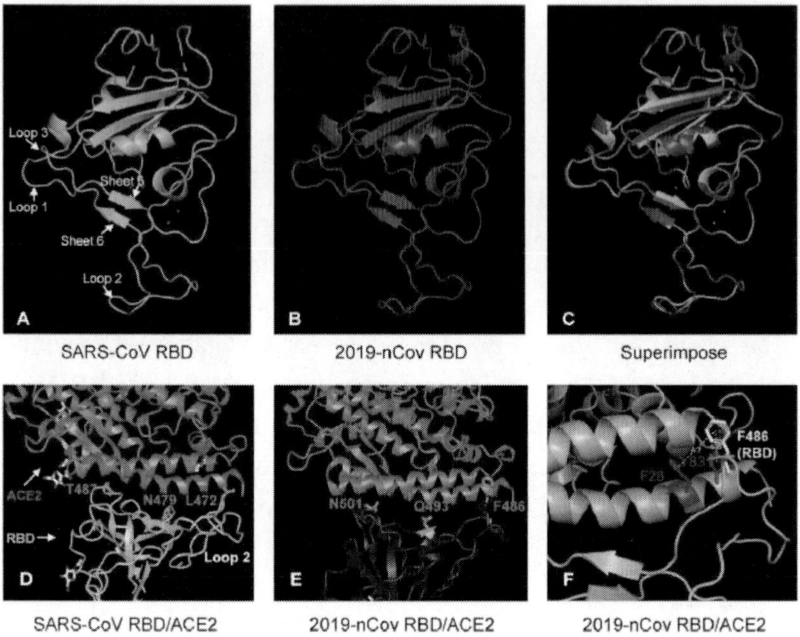

Figure 4. Comparison of the known structures SARS-CoV RBD and its ACE2 complex with the deduced molecular model of 2019-nCoV RBD. Some structure segments and key amino acid residue are indicated (Chen et al., 2020).

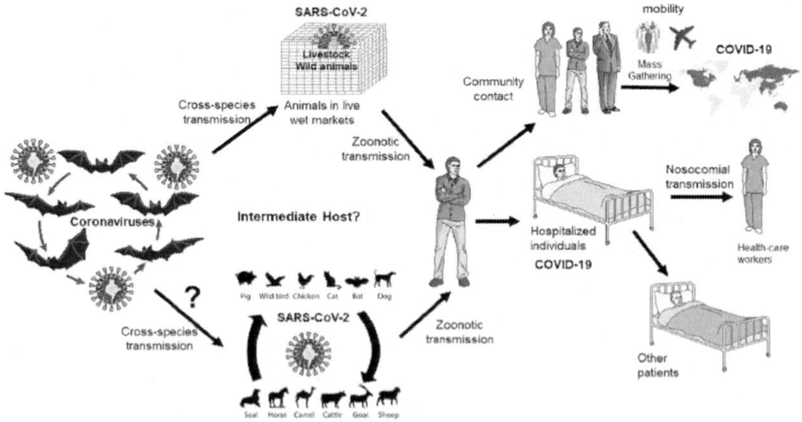

Figure 5. The emergence of SARS-CoV-2 and the outbreak of COVID-19. The figure depicts a hypothesized origin of the virus and a generalized route of transmission of the epidemic zoonotic coronaviurs (El Zowalaty and Jarhult, 2020).

Table 1. Possible origins of SARS-CoV-2

1. The virus evolved to its current pathogenic state via natural selection in a non-human host and then jumped to humans. Bats are more likely reservoir of SARS-Cov-2 because it is similar to a bat coronavirus, and it may be an intermediate host was involved between bats and humans.
2. A non-pathogenic version of the virus transferred from an animal host into humans and then evolved to its current pathogenic state within the human population. Armadillo-like mammals in Asia and Africa, have an RBD structure which is similar to that of SARS-CoV-2. A coronavirus from a pangolin may possibly have been transmitted to a human, directly or through an intermediary host like civets or ferrets.

Andersen et al., 2020.

Table 2. Homologous analysis of SARS-CoV-2 (NC_045512) and six other Coronavirus strains isolated from different hosts in China (%)

Isolate	Host	Complete genome	ORF1ab	N	S
SARS coronavirus civet020 (AY572038)	Civet	73.58	79.23	87.79	71.41
Bat SARS-like coronavirus As6526 (KY417142)	Aselliscus stoliczkanus	74.58	79.23	87.55	68.17
Bat SARS-like coronavirus Rs4874 (KY417150)	Rhinolophus sinicus	71.98	79.18	87.94	71.29
Alphacoronavirus Mink/China/1/2016 (MF113046)	Mink	34.97	38.47	33.70	30.89
Bat coronavirus isolate RaTG13 (MN996532)	Rhinolophus affinis	93.7	96.5	96.9	92.86
Pangolin coronavirus (MT084071)	Manis javanica	?	?	95	90

Note: N, N protein. S, spike protein. ?, Sequence of Pangolin coronavirus (MT084071) is not completed in this part of genome.

Li et al., 2020.

In SARS-CoV-2, M protein is responsible for the transmembrane transport of nutrient, the bud release and the formation of envelope, S protein, attaching to hose receptor ACE2, including two subunits S1 and S2; S1 determines the virus host range and cellular tropism by RBD, and S2 mediates virus-cell membrane fusion by HR1 and HR2. N, E protein and several accessory proteins, interfered with host immune response or unknown function. Li et al., (2020) reported that genome and ORF1a homology show that the virus is not the same coronavirus as the coronavirus derived from five wild animals, namely *Paguma larvata*, *Paradoxurus hermaphrodites*, Civet, *Aselliscus stoliczkanus* and *Rhinolophus sinicus*, whereas the virus has the highest homology with Bat Coronavirus isolate RaTG13. Homologous analysis of SARS-CoV-2 and six other Coronavirus stains isolated from different hosts in China are shown in Table 2. The most important recommendations for travelers to infected countries are presented in Table 3. Populations influenced by COVID-19 divided into 4 levels is shown in Table 4.

Table 3. The most important recommendations for travelers to infected countries

1. Avoid or delay in travel to the cities and areas mainly infected by new virus.
2. Avoid visiting markets which live animals are traded.
3. The consumption of raw or undercooked animal products should be avoided. Food safety practices for raw meat, milk or animal organs should be considered.
4. Avoid large concentrations of people in public areas.
5. Avoid touching objects such as handrails, doorknobs and bringing hands to nose or mouth.
6. Avoid close contact with anyone who has fever and cough.
7. Avoid contact with sick people.
8. Avoid contact with animals both alive or dead.
9. Dispose tissues immediately and wash hands after using.
10.Wash hands frequently with soap and water for at least 20 seconds, or use an alcohol-based hand sanitizer if soap and water are not available.
11.Seek prompt consultation in case of fever and respiratory symptoms during or after the trip.

Table 4. Populations influenced by COVID-19 divided into 4 levels

1. Patients with severe symptoms of COVID-19, front-line medical staff, researchers or administrative staff.
2. Patients with mild symptoms of COVID-19, close contacts, suspected patients or patients with fever who come to hospital for treatment.
3. People related to the first and second-level population, such as family members, colleagues or friends; rescuers, such as commanders, administrative staff, or volunteers.
4. People in affected areas, susceptible groups or general public.

Jiang et al., 2020.

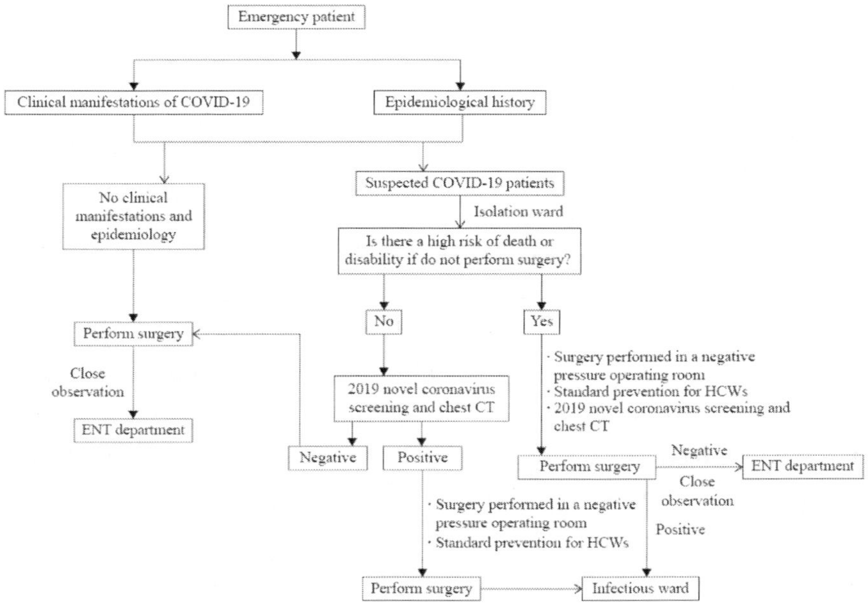

Figure 6. Pathway for the management of emergency ear, nose and throat (ENT) patients. COVID-19, coronavirus disease 2019; CT: computed tomography; HCW: healthcare worker (Lu et al., 2020).

Du et al., (2020) recommended traditional Chinese medicine as an effective treatment for coronavirus 2019 pneumonia. Long et al., (2020) concluded that real-time reverse-transcriptase-polymerase chain reaction (rRT-PCR) may produce initial false negative results, and they have suggested that patients with typical computed tomography (CT) findings but negative rRT-PCR results should be isolated, and rRT-PCR should be repeated to avoid misdiagnosis. Other scientists also stated that final diagnosis relies on rRT-PCR positively for the presence of coronavirus (Corman et al., 2020; Rello et al., 2020; Rubin et al., 2020). However, there are cuurently no effective specific antivirals or drugs combinations supported by high-level evidence (Pan et al., 2020). Wang et al., (2020) introduced spiral chest computed tomography (CT) as a sensitive examination method, which can be applied to make early diagnosis and for evaluation of progression with a diagnostic sensitivity and accuracy better than that of nucleic acid detection. Liu et al., (2020) indicated that on the resolutive phase of the disease, CT abnormalities showed complete resolution or demonstrated residual linear opacities. Perm et al., (2020) noticed that restrictions on activities in Wuhan, would help to delay the epidemic peak and prevent the secondary peak. Sun et al., (2020) highlighted the importance and availability of public datasets to encourage analytical efforts and provide robust evidence to guide interventions. Kobayashi et al., (2020) found that the risk of death among young adults is higher than that of seasonal influenza, and those elderly with underlying comorbidities need additional care. Grubaugh et al., (2019) found that many more mutations will appear in the viral genome which these mutations may help scientists to track the spread of SARS-CoV-2. Coronaviruses have capacity to jump species boundaries and adapt to new hosts (Zhang and Holmes, 2020). Zhao et al., (2020) reported that the majority of patients with suspected or confirmed COVID 19 showed fever and dry cough and presented bilateral multiple mottling and ground-glass opacity on chest computed tomography scans. Lippi et al., (2020) showed that low platelet count is associated with increased risk of severe disease and mortality in patients with COVID-19, and it can be consider as clinical indicator of worsening illness during hospitalization. Lv et al., (2020) emphasized at the importance of successive sampling and testing SARS-Cov-2 by RT-PCR. Xie et al., (2020) recommended combining the computed tomography scans and nucleic acid detection. Guo et al., (2020) indicated the strong influence of COVID-19 epidemic on the utilization or emergency dental services. It has been reported that remdesivir only and in combination with chloroquine or interferon beta significantly blocked the SARS-CoV-2 replication and patients were declared as clinically recovered (Holshue et al., 2020; Sheahan et al., 2020; Wang et al., 2020). Some other anti-virals like Nafamostat, Nitazoxanide, Ribavirin, Penciclovir, Favipiravir, Ritonavir, AAK1, Baricitinib, and Arbidol showed moderate results when tested against infection in patients and in-vitro clinical isolates (Holshue et al., 2020; Richardson et al., 2020; Sheahan et al., 2020; Wang et al., 2020). Shen et al., (2020) consider isothermal nucleic acid amplification as a highly promising candidate method for detection of coronavirus infection, due to its fundamental advatange in quick procedure time at constant temperature without thermocycler operation.

Table 5. Ten functions of the nCapp diagnosis and treatment system for COVID-19 based on the Internet of Things

Function	Significance of diagnosis and treatment of COVID-19
Online monitoring	Best for online monitoring, identifying COVID-19, and guiding graded diagnosis and treatment
Location tracking	Can be used to locate patients diagnosed with COVID-19 and guide treatment when problems are found
Alarm linkage	Can provide alarms to monitor the probability of COVID-19 and provide a three-linkage response function to guide graded diagnosis and treatment
Command and control	Facilitates the graded diagnosis and consultation of patients with COVID-19
Plan management	Presents management criteria for the graded diagnosis and treatment of patients with COVID-19 that can be set in advance for graded management and timely treatment of confirmed, suspected, and suspicious cases
Security privacy	Conducive to providing a corresponding safety guarantee mechanism for the graded diagnosis and treatment of patients with COVID-19
Remote maintenance	Networked services used for the graded diagnosis and treatment of patients with COVID-19
Online upgrade	Ensure the normal operation of the graded diagnosis and treatment of patients with COVID-19 and provides automatic medical service
Command management	Considered beneficial for experts or managers to deeply investigate of expand the diagnosis and treatment functions based on the massive information collected Guides how to better prevent and control COVID-19
Statistical decision	Considered beneficial for experts or managers in performing statistical analysis based on the data of graded diagnosis and treatment of patients with COVID-19 Summarizes experiences, identifies problems, and proposes solutions

Bai et al., 2020.

Seah and Agrawal (2020) found that the ability of SARS-CoV-2 to infect ocular tissue and its pathogenic mechanisms. Ghinai et al., (2020) classified COVIS-19 into four categorizes: high-risk contacts, medium-high-risk contact, medium-risk contacts, low-risk contacts and no-contacts. Pathway for the management of emergency ear, nose and throat (ENT) patients is shown in Figure 6. Ten functions of the nCapp diagnosis and treatment system for COVID-19 based on the Internet of Things are presented in Table 5. The most important Clinical characteristics and treatment of patients with SARS-Cov2 are indicated in Table 6. The most important Diagnostic Criteria for COIVD-19 are shown in Table 7. Most relevant clinical similarities and differences between SARS-CoV and SARS-CoV-2 are presented in Table 8. Reverse transcription-PCR systems for the detection of SARS-CoV-2 RNA from respiratory samples is shown in Figure 7. Alignment of RBM region of S proteins from SARS-CoV-2 and SARS-CoV is indicated in Figure 8. Betacoronaviruses genome organization; The Betacoronavirus for human (SARS-CoV-2, SARS-CoV, and MERS-CoV) genome comprises are shown in Figure 9. Amino Acid substitutions of 2019-nCoV against SARS and SARS-like viruses are indicated in Figure 10.

Table 6. The most important Clinical characteristics and treatment of patients with SARS-Cov2

Signs and symptoms at admission	Chest x-ray and CT findings	Treatment
Fever	Unilateral pneumonia	Oxygen therapy
Cough	Bilateral pneumonia	Mechanical ventilation (Non-invasive, invasive)
Shortness of breath	Multiple mottling and ground-glass opacity	Continuous renal replacement therapy (CRRT)
Muscle ache		Extracorporeal membrane oxygenation (ECMO)
Confusion		Antibiotic treatment
Headache		Antifungal treatment
Sore throat		Glucocorticoids
Rhinorrhoea		Intravenous immunoglobulin therapy
Chest pain		
Diarrhoea		
Nausea and vomiting		

Table 7. The most important Diagnostic Criteria for COIVD-19

Supportive epidemiological history.
1. Clinical manifestation: Fever; normal or low levels of white blood cells or decreased lymphocyte counts at onset. Chest radiology at early stage is characteristic of multiple small patchy shadows and interstitial changes, more prominent in the extrapulmonary bands. Multiple ground-glass opacities and infiltrations may develop bilaterally with disease progression, with possible consolidation in severe cases.
2. Diagnosis: SARS-CoV-2 nucleic acid positive in samples of sputum, pharynx swabs, and secretions of lower respiratory tract tested by real-time reverse-transcriptase-polymerase-chain reaction (rRT-PCR) assay.
3. For patients with acute fever (>37.5oC within 72 hours) and normal chest imaging, if the absolute count of peripheral lymphocytes is less than 0.8×109/L, or the count of DC4+ and CD8+ T cells decrease significantly, isolation and close observation should be conducted at home even it the first SARS-CoV-2 nucleic acid test is negative. Repeat of rRT-PCR should be considered after 24 h, and a chest CT scan should be performed when necessary.

Li, 2020.

Table 8. Most relevant clinical similarities and differences between SARS-CoV and SARS-CoV-2

Characteristic	SARS-CoV	SARS-CoV-2
Target receptor	ACE-2	ACE-2
N protein	IFN-γ inhibitor	Unknown
R0	0.4	1.4-2.5
Chest X-ray	Ground glass opacities	Bilateral, multilobar ground glass opacities
Chest CT-scan	Lobar consolidation Nodular opacities	No nodular opacities
Prevention	Hand hygiene, cough etiquette	Possibly hand hygiene, cough etiquette
Transmission	Droplets Contact with infected individuals	Droplets even asymptomatic ones
Case fatality rate (overall)	9.6%	2.3%

Abbreviations: SARS-CoV= Severe Acute Respiratory Syndrome Coronavirus; SARS-CoV-2= Severe Acute Respiratory Syndrome Coronavirus 2; N protein= Nucleocapsid protein: IFN-γ= interferon-γ; R0= R through; X-ray= radiography; CT-scan= computerized tomography.

Ceccarelli et al., 2020.

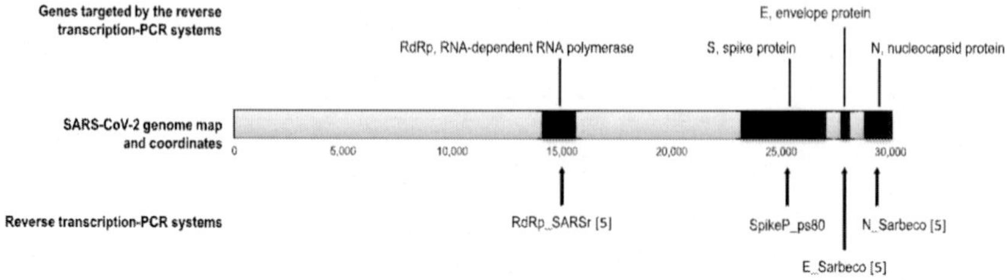

Figure 7. Reverse transcription-PCR systems for the detection of SARS-CoV-2 RNA from respiratory samples (Lagier et al., 2020).

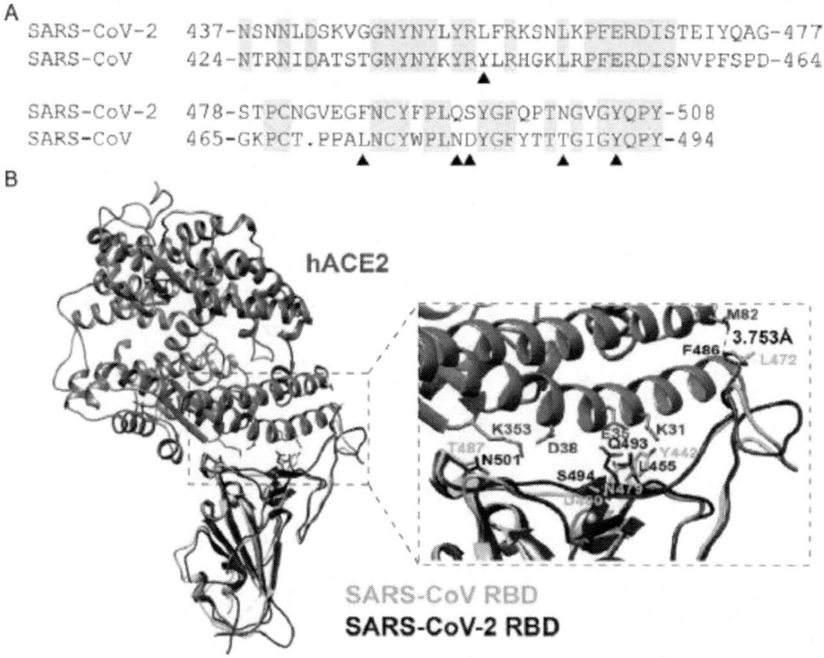

Figure 8. Alignment of RBM region of S proteins from SARS-CoV-2 and SARS-CoV. (A) Sequence alignment of RBM region of S protein from SARS-CoV-2 and SARS-CoV, represents the six key amino acids in the S protein interacting with human ACE2. For SARS-CoV, they are Y442, L472, N479, D480, T487, and Y491. The S protein sequence of SARS-CoV-2 comes from YP_009724390.1, and the S protein sequence of SARS-CoV comes from NP_828851.1. (B) Alignment of the structure of ACE2 recognition of RBD from SARS-CoV-2 and SARS-CoV. Human ACE2 (hACE2), SARS-CoV-2 RBD, and SARS-CoV RBD are in orange red, blue, and green, respectively (Luan et al., 2020).

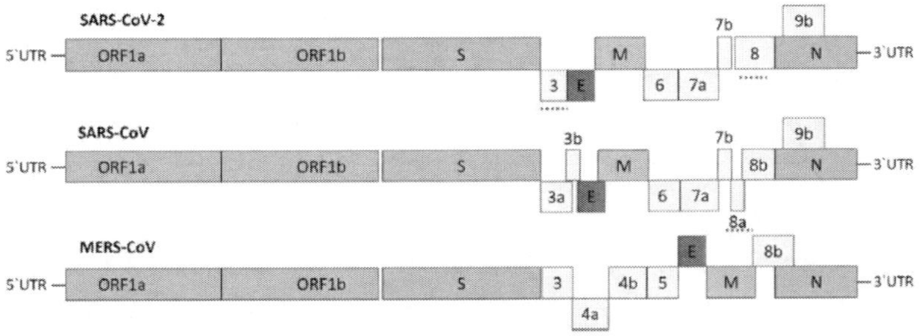

Figure 9. Betacoronaviruses genome organization; The Betacoronavirus for human (SARS-CoV-2, SARS-CoV, and MERS-CoV) genome comprises of the 5/-untranslated region (5/-UTR), open reading frame (orf) 1a/b (green box) encoding non-structural proteins (nsp) for replication, structural proteins including spike (blue box), envelop (maroon box), membrane (pink box), and nucleocapsid (cyan box) proteins, accessory proteins (light gray boxes) such as 3, 6, 7a, 8, and 9b in the SARS-CoV-2 genome, and the 3/-untranslated region (3/-UTR). The doted underlined in red are the protein which shows key variation between SARS-CoV-2 and SARS-CoV. The length of nsps and orfs are not drawn in scale (Shereen et al., 2020).

Table 9. The 26 Chinese herbals screened for COVID-19

Herbal name	The number of antiviral natural compounds in the plant
Forsythiae fructus	3
Licorice	3
Mori cortex	3
Chrysanthemi flos	2
Farfarae flos	2
Lonicerae japonicae flos	2
Mori follum	2
Peucedani radix	2
Rhizoma fagopyri cymosi	2
Tmaricis cacumen	3
Erigeron breviscapus	2
Radix bupleuri	2
Coptidis rhizome	2
Houttuyniae herba	2
Hoveniae dulcis semen	2
Inulae flos	2
Eriobotryae folium	3
Hedysarum multijugum maxim.	3
Lepidii semen descurainiae semen	3

Table 9. (Continued)

Herbal name	The number of antiviral natural compounds in the plant
Ardisiae japonicae herba	2
Asteris radix et rhizome	2
Euphorbiae helioscopiae herba	2
Ginkgo semen	2
Anemarrhenae rhizome	3
Epimrdii herba	2
Fortunes bossfern rhizome	2

Zhang et al., 2020.

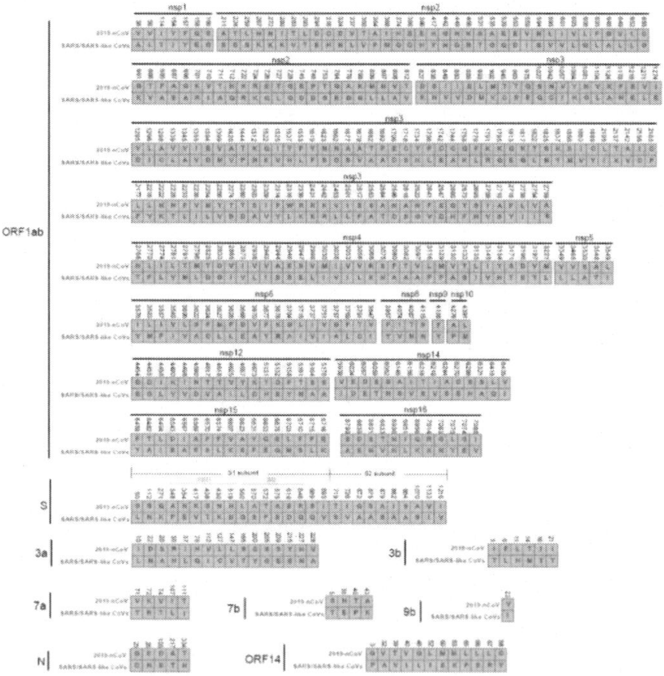

Figure 10. Amino Acid substitutions of 2019-nCoV against SARS and SARS-like viruses. All 27 proteins encoded by 2019-nCoV have been aligned against SARS-CoVs and SARS-like bat CoVs using the FFT-NS-2 algorithm in MAFFT (version v7.407). An amino acid substitution was defined as an absolutely conserved site in the group of SARS and SARS-like CoVs but different from that of 2019-nCoV. In total, 380 amino acid substitutions have been identified between the amino acid sequences of 2019-nCoV (HB01) and the corresponding consensus sequences of SARS and SARS-like CoVs (Wu et al., 2020).

Zhang et al. (2020) found that 26 herbals plants which have significant roles to regulate viral infection, immune/inflammation reactions and hypoxia response. The 26 Chinese herbals screened for COVID-19 is presented in Table 9.

CONCLUSION

Severe acute respiratory syndrome (SARS) is caused by a coronavirus SARS-CoV which was started in pigs or ducks in south of China and mutated to affect humans. It was originated in Guangdong province. CoVs are single-stranded RNA viruses which belong to the order Nidovirales, family Coronaviridae, and subfamily Coronavirinae. The viruses are subdivided into four genera on the basis of genotypic and serological characters which are Alpha-, Beta-, Gamma, and Deltacoronavirus, and among them the first two genera are those which infect humans. Seven coronaviruses are known to infect humans, three of them are serious, namely, SARS (severe acute respiratory syndrome, China, 2002), MERS (Middle East respiratory syndrome, Saudi Arabia, 2012), and SARS-CoV-2 (2019-2020). SARS is caused by a coronavirus (SARS-CoV) which exists in bats and palm civets in Southern China. Its family is *Coronaviridae*, and its genus is *Coronavirus*. It is enveloped, helical nucleocapsid, spherical to pleomorphic, helical nucleocapsid, spherical to pleomorphic, kidney-shaped or rod-shaped particles, 100-130 nm in diameter. Its nucleaic acid is linear, positive-sense, single-stranded RNA, ~29.8 kb in length. Virions sensitive to treatment with lipid solvents, nonionic detergents, formaldehyde, and oxidizing agents. The most important groups who are at risk are family members in close contact with cases, health-care workers in close contact with cases, elderly and immune compromised individuals appear at increased risk. During the first week, infected people have symptoms of a flu-like illness, followed by atypical pneumonia such as dry cough, and progressively worsening shortness of breath with poor oxygenation. During its outbreak, nearly 25% of people had severe respiratory failure and 10% died (case fatality rates range from 6.6 to 17%), and it was controlled by using public-health measures, namely, wearing surgical masks, washing hands and isolating infected patients. The civet cat (palm civet) in Southeast Asia is likely source of introduction of the agent to humans. The SARS CoV RNA sequence found in palm civets is 99% identical to that found in plam civets is 99% identical to that found in humans. Similar RNA sequences have been found in bats, snakes and monkeys. No medication has been proven to treat SARS effectively, but oxygen therapy and tracheal intubation and mechanical ventilation to support life until recovery begins is useful for patients in severe case. The most useful ways to control SARS pandemic is public-health and infection-control measures. The most important primary measures are isolation, ribavirin, and corticosteroid therapy, mechanical ventilation, convalescent plasma, and others. It can be spread from close person-to-person contact via respiratory droplets which come in contact with skin or mucous membranes such as eyes, mouth or nose.

MERS-CoV is a zoonotic virus which can lead to secondary human infections. Dromedary camels have been recognized as the intermediate host, with closely related virus sequences in bats. MERS carries a 35% mortality rate. There is no clear and specific treatment for MERS, and person to person spread causes hospital and household outbreaks of MERS-CoV. It is the sixth coronavirus that influences human. Like other coronaviruses, it is an enveloped single-stranded RNA virus which replicates in the host-cell cytoplasm, with approximate size of 30 kb. It has structural proteins, called the E, M, and N proteins, and membrane protein called the spike (S) protein, which has important role in the virus attachment and entry into the host cells. It was first discovered in September 2012 as the cause of death in patients who had died of severe pneumonia in June 2012 in Jeddah, Saudi Arabia. MERS-CoV is most likely derived from an ancestral reservoir bats. MERS outbreak was found in the Republic of Korea since 2015. Coronavirus entry is initiated by the binding of the spike protein (S) to cell receptors, specifically, dipeptidyl peptidase 4 (DDP4) and angiotensin converting enzyme 2 (ACE2) for MERS-CoV and SARS-CoV, respectively. It has been concluded that the MERS-S protein is known to represent a key target for the development of new therapeutics and includes of a receptor-binding subunit S1 and a membrane-fusion subunit S2. The subunit S1 is composed of four different core domain, and the domain S1B binds to the host-cell receptor dipeptidylpeptidase 4 (DPP4), while the domain S1A binds to sialoglycans which increased infection of human lung cells by MERS-CoV. The roles of S protein in receptor binding and membrane fusion make it a perfect target for vaccine and antiviral development. It has been showed that vaccines based on the S protein can induce antibodies to block virus binding and fusion or neutralize virus infection. The most important treatment for coronavirus is ribavirin/interferon combination, lopinavir-ritonavir, convalescent plasma, monoclonal antibodies, polyclonal antibodies, mycophenolate mofetil, and remdesivir. Successful vaccine should be balanced to achieve protection.

The novel SARS-CoV-2 coronavirus which was started in the city of Wuhan, spread in all over the world. Its genome sequence analysis has shown that SARS-CoV-2 belons to betacoronavirus genus, which includes Bat SARS-like coronavirus, SARS-CoV and MERS-CoV. On the basis of nucleic acid sequence similarity, the newly identified 2019-nCoV is a betacoronavirus. The RBD portion of the SARS-CoV-2 pike proteins has evolved to effectively target a molecular feature on the outside of human cells called ACE2, a receptor involved in regulating blood pressure. The SARS-CoV-2 spike protein was found so effective at binding the human cells. In SARS-CoV-2, M protein is responsible for the transmembrane transport of nutrient, the bud release and the formation of envelope, S protein, attaching to hose receptor ACE2, including two subunits S1 and S2. It has been reported that remdesivir only and in combination with chloroquine or interferon beta significantly blocked the SARS-CoV-2 replication and patients were declared as clinically recovered. Some other anti-virals like Nafamostat, Nitazoxanide, Ribavirin, Penciclovir,

Favipiravir, Ritonavir, AAK1, Baricitinib, and Arbidol showed moderate results when tested against infection in patients and in-vitro clinical isolates. Isothermal nucleic acid amplification as a highly promising candidate method for detection of coronavirus infection, due to its fundamental advatange in quick procedure time at constant temperature without thermocycler operation. These diseases can be considered important models for emerging infectious diseases as it emerged from natural animal reservoirs. Early recognition, prompt isolation and appropriate supportive therapy are the main parameters in combating with these deadly infections.

REFERENCES

Abduljali, J. M., & Abduljali, B. M. (2020). Epidemiology, genome and clinical features of the pandemic SARS-CoV-2: a recent view. *New Microbes and new Infections*. http://doi.org/10.1016/j.nmni.2020.100672.

Adams, M. J., & Carstens, E. B. (2012). Ratification vote on taxonomic proposals to the International Committee on Taxonomy of Viruses. *Arch Virl*, 157, 1411-1422.

Adegboye, O., Saffary, T., Adegboye, M., & Elfaki, F. (2019). Individual and network characteristics associated with hospital-acquired Middle East Respiratory Syndrome coronavirus. *Journal of Infection and Public Health*, 12, 343-349.

Agostini, M. L., Pruijssers, A. J., & Chappell, J. D., et al., (2019). Small-molecule antiviral beta-d-n^4-hydroxycytidine inhibits a proofreading-intact coronavirus with a high genetic barrier to resistance. *J Virol*, 93, e01348-19.

Ahmad, A., Krumkamp, R., & Reintjes, R. (2009). Controlling SARS: a review on China's response compared with other SARS-affected countries. *Tropical Medicine and International Health*, 14(1), 36-45.

Ahmadzadeh, J., Mobaraki, K., Mousavi, S. J., Aghazadeh-Attari, J., Mirza-Aghazadeh-Attari, M., & Mohebbi, I. (2020). The risk factors associated with MERS-CoV patient fatality: A global survey. *Diagnostic Microbiology and Infectious*, 96, 114876.

Ahmed, A. E. (2017). Diagnostic delays 537 symptomatic cases of Middle East respiratory syndrome coronavirus infection in Saudi Arabia. *Int J Infect Dis*, 62(September), 47-51.

Ahmed, A. E. (2019). Diagnostic delays in Middle East respiratory syndrome coronavirus patients and health systems. *Journal of Infection and Public Health*, 12, 767-771.

Akerstrom, S., Gunalan, V., Tat Keng, C., Tan, Y. J., & Mirazimi, A. (2009). Dual effect of nitric oxide on SARS-CoV replication: Viral RNA production and palmitoylation of the S protein are affected. *Virology*, 3950, 1-9.

Al-Khannaq, M. N., Ng, K. T., Oong, X. Y., Pang, Y. K., Takebe, Y., Chook, J. B., Hanafi, N. S., Kamarulzaman, A., & Tee, K. K. (2016). Molecular epidemiology and evolutionary histories of human coronavirus OC43 and HKU1 among patients with upper respiratory tract infections in Kuala Lumpur, Malaysia. *Virology Journal*, 13, 33.

Alagaili, A. N., Briese, T., Amor, N. M. S., Mohammed, O. B., & Lipkin, W. I. (2019). Waterpipe smoking as a public health risk: potential risk for transmission of MERS-CoV. *Saudi Journal of Biological Sciences*, 26, 938-941.

Alfaraj, S. H., Al-Tawfiq, J. A., Assiri, A. Y., Alzahrani, N. A., Alanazi, A. A., & Memish, Z. A. (2019). Clinical predictors of mortality of Middle East Respiratory Syndrome Coronavirus (MERS-CoV) infection: a cohort study. *Travel Medicine and Infectious Disease*, 29, 48-50.

Alfaraj, S. H., Al-Tawfiq, J. A., Memish, Z. A., FRCPC, FACP, FRCPE & FRCPL. (2019). Middle East respiratory syndrome coronavirus intermittent positive cases: Implications for infection control. *American Journal of Infection Control*, 47, 290-293.

Alharbi, N. K., Padron-Regalado, E., Thompson, C. P., Kupke, A., Wells, D., & Sloan, M. A., et al., (2017). ChAdOx1 and MVA based vaccine candidates against MERS-CoV elicit neutralizing antibodies and cellular immune responses in mice. *Vaccine*, 35, 3780-3788.

Al-Jasser, F. S., Nouh, R. M., & Youssef, R. M. (2019). Epidemiology and predictors of survival of MERS-CoV infections in Riyadh region, 2014-2015. *Journal of Infection and Public Health*, 12, 171-177.

Almaghrabi, R. S., & Omrani, A. S. (2017). Middle East respiratory syndrome coronavirus (MERS-CoV) infection. *Br J Hosp Med* (Lond), 78(1), 23-26.

Al-Mohrej, O. A., Al-Shirian, S. D., Al-Otaibi, S. K., Tamim, H. M., Masuadi, E. M., & Fakhoury, H. M. (2016). Is the Saudi public aware of Middle East respiratory syndrome? *J Infect Public Health*, 9(3), 259-266.

Almutairi, K. M., Al Helih, E. M., Moussa, M., Boshaigah, A. E., Saleh Alajilan, A., & Vinluan, J. M., et al., (2015). Awareness, attitude, and practices related to coronavirus pandemic among public in Saudi Arabia. *Farm Community Health*, 38(4), 332-340.

Al-Omari, A., Rabaan, A. A., Salih, S., Al-Tawfiq, J. A., & Memish, Z. A. (2019). MERS coronavirus outbreak: Implications for emerging viral infections. *Diagnostic Microbiology and Infectious Disease*, 93, 265-285.

Al-Tawfiq, J. A. (2013). Middle East Respiratory Syndrome-coronavirus infection: An overview. *Journal of Infection and Public Health*, 6, 319-322.

Al-Tawfiq, J. A., & Memish, Z. A. (2014). Emerging respiratory viral infections: MERS-CoV and influenza. *Lancent Respir Med*, 2(1), 23-25.

Al-Tawfiq, J. A., Zumla, A., & Memish, Z. A. (2014). Travel implications of emerging coronaviruses: SARS and MERS-CoV. *Travel Medicine and Infectious Disease*, 12, 422-428.

Al-Tawfiq, J. A., & Pear, T. M. (2015). Middle East respiratory syndrome coronavirus in healthcare settings. *Curr Opin Infect Dis*, 28, 392-396.

Al-Tawfiq, J. A., & Memish, Z. A. (2016). Drivers of MERS-CoV transmission: what do we know? *Expert Rev Respir Med*, 10, 331-338.

Al-Tawfiq, J. A., Rabaan, A. A., & Hinedi, K. (2017). Influenza is more common than Middle East respiratory syndrome coronavirus (MERS-

CoV) among hospitalized adult Saudi patients. *Travel Medicine and Infectious Disease*, 20, 56-60.

Al-Tawfiq, J. A., & Auwaerter, P. G. (2019). Healthcare-associated infections: the hallmark of Middle East respiratory syndrome coronavirus with review of the literature. *Journal of Hospital Infection*, 101, 20-29.

Al-Tawfiq, J. A., Abdrabalnabi, R., Taher, A., Mathew, S., & Rahman, K. A. (2019). Infection control influence of Middle East respiratory syndrome coronavirus: A hospital-based analysis. *American Journal of Infection Control*, 47, 431-434.

Alqahtani, A. S., Rashid, H., Basyouni, M. H., Alhawassi, T. M., & BinDhim, N. F. (2017). Public response to MERS-CoV in the Middle East: iPhone survey in six countries. *Journal of Infection and Public Health*, 10, 534-540.

Amer, H., Alqahtani, A. S., Alaklobi, F., Altayeb, J., & Memish, Z. A. (2018). Healthcare worker exposure to Middle East respiratory syndrome coronavirus (MERS-CoV): Revision of screening strategies urgently needed. I*nternational Journal of Infectious Diseases*, 71, 113-116.

Andersen, K. G., Rambaut, A., Lipkin, W. I., Holmes, E. C., & Garry, R. F. (2020). The proximal origin of SARS-CoV-2. *Nature Medicine*. doi: 10.1038/s41591-020-0820-9.

Ar Gouilh, M., Puechmaille, S. J., Diancourt, L., Vandenbogaert, M., Serra-Cobo, J., Roig, M. L., Brown, P., Moutou, F., Caro, V., Vabret, A., & Manuguerra, J. C., et al., (2018). SARS-CoV related Betacoronavirus and diverse Alphacoronavirus members found in western old-world. *Virology*, 517, 88-97.

Arabi, Y. M., Arifi, A. A., Kalkhy, H. H., Najm, H., Aldawwod, A. S., & Ghabashi, A., et al., (2014). Clinical course and outcomes of critically ill patients with Middle East respiratory syndrome coronavirus infection. *Ann Intern Med*, 160, 389-397.

Arabi, Y., Balkhy, H., & Hajeer, A. H., et al., (2015). Feasibility, safety, clinical, and laboratory effects of convalescent plasma therapy for patients with Middle East respiratory syndrome coronavirus infection: a study protocol. *Springerplus*, 4, 709.

Arabi, Y. M., Alothman, A., & Balkhy, H. H., et al., (2018). Treatment of Middle East respiratory syndrome with a combination of lopinavir-ritonavir and interferon-beta1b (MIRACLE trial): study protocol for a randomized controlled trial. *Trials*, 19, 81.

Aronin, S. I., & Sadigh, M. (2004). Severe acute respiratory syndrome. *Conn Med*, 68(4), 207-215.

Assiri, McGeer, A., Perl, T. M., Price, C. S., Al Rabeeah, A. A., Cummings, D. A., Alabdullatid, Z. N., Assad, M., Almulhim, A., Makhdoom, H., Madani, H., Alhakeem, R., Al-Tawfig, J. A., Cotton, M., Watson, S. J., Kellam, P., Zumla, A. I., & Memish, Z. A. (2013). Hospital outbreak or Middle East respiratory syndrome coronavirus. *N Engl J Med*, 369(5), 407-416.

Assiri, A., Al-Tawfiq, J. A., Al-Rabeeah, A. A., Al-Rabiah, F. A., Al-Hajjar, S., & Al-Barrak, A., et al., (2013). Epidemiological, demographic, and clinical characteristics of 47 cases of Middle East respiratory syndrome coronavirus disease from Saudi Arabia: a descriptive study. *Lancent Infect Dis*, 13, 752-761.

Assiri, A., Abedi, G. R., Saeed, A. A. B., Abdalla, M. A., al-Masry, M., & Choudhry et al., (2016). Multifacility outbreak of Middle East respiratory syndrome in Tarif. *Saudi Arabia Emerg Infect Dis*, 22, 32-40.

Badawi, A., & Ryoo, S. G. (2016). Prevalence of comorbidities in the Middle East respiratory syndrome coronavirus (MERS-CoV): a systematic review and meta-analysis. *International Journal of Infectious Diseases*, 49, 129-133.

Baharoon, S., & Memish, Z. A. (2019). MERS-CoV as an emerging respiratory illness: A review of prevention methods. *Travel Medicine and Infectious Disease*, 32, 101520.

Bakkers, M. J. G., Lang, Y., Feitsma, L. J., Hulswit, R. J. G., de Poot, S. A. H., van Wliet, A. L. W., Margine, I., de Groot-Mijnes, J. D. F., van Kuppeveld, F. J. M., Langereis, M. A., Huizinga, E. G., & de Groot, R. J. (2017). Betacoronavirus adaptation to human involved progressive loss of hemagglutinin-esterase lectin activity. *Cell Host and Microbe*, 21, 356-366.

Banach, B. S., Orenstein, J. M., Fox, L. M., Randell, S. H., Rowley, A. H., & Baker, S. C. (2009). Human airway epithelial cell culture to identify new respiratory viruses: Coronavirus NL63 as a model. *Journal of Virological Methods*, 156, 19-26.

Banik, G. R., Khandaker, G., & Rashid, H. (2015). Middle East Respiratory Syndrome Coronavirus "MERS-CoV": current knowledge gaps. *Paediatric Respiratory Reviews*, 16, 197-202.

Bai, L., Yang, D., Wang, X., Tong, L., Zhu, X., Zhong, N., & Bai, C., et al., (2020). Chinese experts' consensus on the Internet of Things-aided diagnosis and treatment of coronavirus disease 2019 (COVID-19). *Clinical eHealth*, 3, 7-15.

Bal, A., Destras, G., Gaymard, A., Bouscambert-Duchamp, M., Valette, M., Escuret, V., Frobert, E., Billaud, G., Trouillet-Assant, S., Cheynet, V., Brengel-Pesce, K., Morfin, F., Lina, B., & Josset, L. (2020). Molecular characterization of SARS-CoV-2 in the first COVID-19 cluster in France reveals an amino-acid deletion in nsp2 (Asp268Del). *Clinical Microbiology and Infection*. http://doi.org/10.1016/j.cmi.2020.03.020

Bawazir, A., Al-Mazroo, E., Jradi, H., Ahmed, A., & Badir, M. (2018). MERS-CoV infection: Mind the public knowledge gap. *Journal of Infection and Public Health*, 11, 89-93.

Behzadi, M. A., & Leyva-Grado, V. H. (2019). Overview of current therapeutics and novel candidates against influenza, respiratory syncytial virus, and Middle East respiratory syndrome coronavirus infections. *Front Microiol*, 10, 1327.

Beidas, M., & Chehadeh, W. (2018). Effect of human coronavirus OC43 structural and accessory proteins on the transcriptional activation of antiviral response elements. *Intervirology*, 61, 30-35.

Beigel, J. H., Nam, H. H., & Adams, P. L., et al. (2019). Advances in respiratory virus therapeutics- a meeting report from the 6th ISIRV antiviral group conference. *Antiviral Res*, 167, 45-67.

Benvenuto, D., Giovanetti, M., Vassallo, L., Angeletti, S., & Ciccozzi, M. (2020). Application of the ARIMA model on the COVID-2019 epidemic dataset. *Data in Brief*. 29: 105340.

Bermingham, A., Chand, M. A., Brown, C. S., Aarons, E., Tong, C., Langrish, C., & Zambon, M. (2012). Severe respiratory illness caused by a novel coronavirus, in a patient transderred to the United Kingdom from the Middle East, September 2012. *Euro Surveill*. 17(40): 20290.

Bogoch, I. I., Creatore, M. I., Cetron, M. S., Brownstein, J. S., Pesik, N., Miniota, J., Tam, T., Hu, W., Nicolucci, A., Ahmed, S., Yoon, J. W., Berry, I., Hay, S. I., Anema, A., Tatem, A. J., MacFadden, D., German, M., & Khan, K. (2015). Assessment of the potential for international dissemination of Ebola virus via commercial air travel during the 2014 West African outbreak. *Lancet*. 385(9962): 29-35.

Booth, C. M., Matukas, L. M., Tomlinson, G. A., Rachlis, A. R., Rose, D. B., Dwosh, H. A., Walmsley, S. L., Mazzulli, T., Avendano, M., Derkach, P., Ephtimios, I. E., Kitai, I., Mederski, B. D., Shadowitz, S. B., Gold, W. L., Hawryluck, L. A., Rea, E., Chenkin, J. S., Cescon, D. W., Poutanen, S. M., & Detsky, A. S. (2003). Clinical features and short-term outcomes of 144 patients with SARS in the greater Toronto area. *JAMA*.

Bosch, B. J., van der Zee, R., de Haan, C. A. M., & Rottier, P. J. M. (2003). The coronavirus spike protein is a class I virus fusion protein: structural and functional characterization of the fusion core complex. *J. Virol*. 77: 8801-8811.

Bryce, E., Copes, R., Gamage, B., Lockhart, K., & Yassi, A. (2008). Staff perception and institutional reporting: two views of infection control compliance in British Columbia and Ontario three years after an outbreak of severe acute respiratory syndrome. *Journal of Hospital Infection*. 69: 169-176.

Cao, C., Chen, W., Zheng, S., Zhao, J., Wang, J., & Cao, W. (2016). Analysus of spatiotemporal characteristics of pandemic SARS spread in mainland China. *BioMed Research International*. Volume 2016, Article ID, 12 pages.

Carbajo-Lozoya, J., Ma-Lauer, Y., Malesevic, M., Theuerkorn, M., Kahlert, V., Prell, E., von Brunn, B., Muth, D., Baumert, T. F., Drosten, C., Fischer, G., & von Brunn, A. (2014). Human coronavirus NL63 replication is cyclophilin A-dependent and inhibited by ion-immunosuppressive cyclosporine A-derivatives including Alisporivir. *Virus Research*. 184: 44-53.

Ceccarelli, M., Berretta, M., Venanzi Rullo, E., Nunnari, G., & Cacopardo, B. (2020). Editorial- difference and similarities between severe acute respiratory syndrome (SARS)-Coronavirus (CoV) an SARS-CoV-2. Would a rose by another name smell as sweet? *European Review for Medical and Pharmacological Sciences*. 24: 2781-2783.

Centers for Disease Control and Prevention: *Updated interim US case definition for severe acute respiratory syndrome (SARS).* URL: http://www.cdc.gov/ncidod/sars/casedefinition.htm.

Cha, M. J., Chung, M. J., Kim, K., Lee, K. S., Kim, T. J., & Kim, T. S. (2018). Clinical implication of radiographic scores in acute Middle East respiratory syndrome coronavirus pneumonia: Report from a single tertiary-referral center of South Korea. *European Journal of Radiology*. 107: 196-202.

Chan-Yeung, M., & Yu, W. C. (2003). Outbreak of severe acute respiratory syndrome in Hong Kong special administrative region: case report. *BMJ*. 326: 850-852.

Chan, J. F. W., Lau, S. K. P., To, K. K. W., Cheng, V. C. C., Woo, P. C. Y., & Yuen, K. Y. (2015). Middle East respiratory syndrome coronavirus: another zoonotic betacoronavirus causing SARS-like disease. *Clin Microbiol Rev*. 28(2): 465-522.

Chan, J. F. W., Yuan, S., Kok, K. H., To, K. K. W., Chu, H., Yang, J., Xing, F., Liu, J., Yip, C. C. Y., Poon, R. W. S., Tsoi, H. W., Lo, S. K. F., Chan, K. H., Poon, V. K. M., Chan, W. M., lp, J. D., Cai, J. P., Cheng, V. C. C., Chen, H., Hui, C. L. M., & Yuen, K. Y. (2020). A familial cluster of pneumonia associated with the 2019 novel coronavirus indicating person-to-person transmission: a study of a family cluster. *Lancet*. 395: 514-523.

Chang, L., Yan, Y., & Wang, L. (2020). Coronavirus disease 2019: coronaviruses and blood safety. *Transfusion Medicine Reviews.* http://doi.org/10.1016/j,tmrv.2020.02.003

Chee, Y. C. 2003. Severe acute respiratory syndrome (SARS)-150 days on. *Annals of the Academy of Medicine, Singapore*. 32(3): 277-280.

Chen, Y., Guo, Y., Pan, Y., and Zhao, Z. J. (2020). Structure analysis of the receptor binding of 2019-nCoV. *Biochemical and Biophysical Research Communications*. 525: 135-140.

Cheng, S. C., Chang, Y. C., Chiang, Y. L. F., Chien, Y. C., Cheng, M., Yang, C. H., Huang, C. H., & Hsu, Y. N. (2020). First case of coronavirus disease 2019 (COVID-19) pneumonia in Taiwan. *Journal of the Formosan Medical Association*. 119: 747-751.

Chernomordik, L. V., & Kozlov, M. M. (2008). Mechanics of membrane fusion. *Nat. Struct. Mol. Biol.* 15: 675-683.

Cheung, T. M. T., Yam, L. Y. C., Lau, A. C. W., Kong, B. M. H., & Yung, R. W. H. (2004). Effectiveness of noninvasive positive pressure ventilation in the treatment of acute respiratory failure in severe acute respiratory syndrome. *CHEST*. 126: 845-850.

Chowell, G., Fenimore, P. W., Castillo-Garsow, M. A., & Castillo-Chavez, C. (2003). SARS outbreaks in Ontario, Hong Kong and Singapore: the role of diagnosis and isolation as a control mechanism. *Journal of Theoretical Biology*. 224: 1-8.

Chowell, G., Blumberg, S., Simonsen, L., Miller, M. A., & Viboud, C. (2014). Synthesizing data and models for the spread of MERS-CoV, 2013; key role of index cases and hospital transmission. *Epidemics*. 9: 40-51.

Choudhry, H., Bakhrebah, M. A., Abdulaal, W. H., Zamzami, M. A., Baothman, O. A., Hassan, M. A., Zeyadi, M., Helmi, N., Alzahrani, F. A., Ali, A., Zakaria M. K., Kamal, M. A., Warsi, M. K., Ahmed, F., Rasool, M., & Jamal, M. S. (2019). Middle East respiratory syndrome: pathogenesis and therapeutic developments. *Future Virol*. 14(4): 237-246.

Christian, M. D., Poutanen, S. M., Loutfy, M. R., Muller, M. P., & Low, D. E. (2004). Severe acute respiratory syndrome. *Clinical Infectious Diseases*. 38: 1420-1427.

Chu, K. H., Tsang, W. K., Tang, C. S., Lam, M. F., Lai, F. M., To, K. F., Fung, K. S., Tang, H. L., Yan, W. W., Chan, H. W. H., Lai, T. S. T., Tong, K. L., & Lai, K. N. (2005). Acute renal impairment in coronavirus-associated severe acute respiratory syndrome. *Kidney International*. 67: 698-705.

Chu, Y. K., Ali, G. D., Jia, F., Li, Q., Kelvin, D., Couch, R. C., Harrod, K. S., Hutt, J. A., Cameron, C., Weiss, S. R., & Jonsson, C. B. (2008). The SARS-CoV ferret model in an infection-challenge study. *Virology*. 374: 151-163.

Coleman, P. M., & Lawrence, M. C. (2003). The structural biology of type I viral membrane fusion. *Nat. Rev. Mol. Cell Biol*. 4: 309-319.

Coleman, C. M., & Frieman, M. B. (2014). Coronaviruses: important emerging human pathogens. *J. Virol*. 88: 5209-5212.

Coleman, C. M., Venkataraman, T., Liu, Y. V., Glenn, G. M., Smith, G. E., & Flyer, D. C., et al., (2017). MERS-CoV spike nanoparticles protect mice from MERS-CoV infection. *Vaccine*. 35(12): 1586-1589.

Corman, V. M., Ithete, N. L., Richards, L. R., Schoeman, M. C., Presier, W., Drosten, C., & Drexler, J. F. (2014). Rooting the phylogenetic tree o Middle East respiratory syndrome coronavirus by characterization of a conspecific virus from and African bat. *J. Virol*. 88: 11297-11303.

Corman, V. M., Olschlager, S., Wendtner, C. M., Drexler, J. F., Hess, M., & Drosten, C. (2014). Performance and clinical validation of the RealStar MERS-CoV kit for detection of Middle East respiratory syndrome coronavirus RNA. *Journal of Clinical Virology*. 60: 168-171.

Corman, V. M., Landt, O., & Kaiser, M., et al., (2020). Detection of 2019 novel coronavirus (2019-nCoV) by real-time RT-PCR. *Euro Surveill*. 25(3).

Coutard, B., Valle, C., de Lamballerie, X., Canard, B., Seidah, N. G., & Decroly, E. (2020). The spike glycoprotein of the new coronavirus

2019-nCoV contains a furinlike cleavage site absent in CoV of the same clade. *Antiviral Research*. 176: 104742.

Cowling, B. J., Park, M., Fang, V. J., Wu, P., Leung, G. M., & Wu, J. T. (2015). Preliminary epidemiological assessment of MERS-CoV outbreak in South Korea, May to June 2015. *Eurosurveillance*. 20(25).

Cui, L. J., Zhang, C., Zhang, T., Lu, R. J., Xie, Z. D., Zhang, L. L., Liu, C. Y., Zhou, W. M., Ruan, L., Ma, X. J., & Tab, W. J. (2010). Human coronaviruses HCoV-NL63 and HCoV-HKU1 in hospitalized children with acute respiratory infections in Beijing, China. *Advances in Virology*. Article ID 129134, 6 pages.

Cui, J., Eden, J. S., Holmes, E. C., & Wang, L. F. (2013). Adaptive evolution of bat dipeptidyl peptidase 4 (DPP4): implications for the origin and emergence of Middle East respiratory syndrome coronavirus. *Virol J*. 10: 304-304.

Cunha, C. B., & Opal, S. M. (2014). Middle East respiratory syndrome (MERS). *Virulence*. 5(6): 650-654.

Cyranoski, D. (2020). Mystery deepens over animal source of coronavirus. *Nature*. 579: 18-9. http://doi.org/10.1038/d41586-020-00548-w.

DeDiego, M. L., Alvarez, E., Almazan, F., Rejas, M. T., Lamirande, E., Roberts, A., Shieh, W. J., Zaki, S. R., Subbarao, K., & Enjuanes, L. (2007). A severe acute respiratory syndrome coronavirus that lacks the E gene is attenuated *in vitro* and *in vivo*. *J. Virol*. 81(4): 1701-1713.

De Groot, R. J., Baker, S. C., Baric, R. S., Brown, C. S., Drosten, C., Enjuanes, L., Fouchier, R. A., Galiano, M., Gorbalenya, A. E., Memish, Z. A., Perlman, S., Poon, L. L., Snijder, E. J., Stephens, G. M., Woo, P. C., Zaki, A. M., Zambon, M., & Ziebuhr, J. (2013). Middle East respiratory syndrome coronavirus (MERS-CoV): announcement of the Coronavirus Study Group. *J. Virol*. 87: 7790-7792.

De Wit, E., & Munster, V. J. (2013). MERS-CoV: the intermediate host identified? Lancet *Infect Dis*. 13: 827-828.

De Wit, E., van Doremalen, N., Falzarano, D., & Munster, V. J. (2016). SARS and MERS: recent insights into emerging coronaviruses. *Nat. Rev. Microbiol*. 14: 523-534.

DeDiego, M. L., Pewe, L., Alvarez, E., Rejas, M. T., Perlman, S., & Enjuanes, L. (2008). Pathogenicity of severe acute respiratory coronavirus deletion mutants in hACE-2 transgenic mice. *Virology*. 376(2): 379-389.

DeDiego, M. K., Nieto-Torres, J. L., Jimenez-Guardeno, J. M., Regla-Nava, J. A., Alvarez, E., Oliveros, J. C., Zhao, J., Fett, C., Perlman, S., & Enjuanes, L. (2011). Severe acute respiratory syndrome coronavirus envelope protein regulates cell stress response and apoptosis. *PLOS Pathog*. 7: e1002315.

DeDiego, M. L., Nieto-Torres, J. L., Jimenez-Guardeno, J. M., Regla-Nava, J. A., Castano-Rodriguez, C., Fernandez-Delgado, R., Usera, F., & Enjuanes, L. (2014). Coronavirus virulence genes with main focus on SARS-CoV envelope gene. *Virus Research*. 194: 124-137.

Demmler, G. J., & Ligon, B. L. (2003). Severe acute respiratory syndrome (SARS): a review of the history, epidemiology, prevention and concerns for the future. *Seminars in Pediatric Infectious Diseases*. 1493): 240-244.

Derrick, J. L., & Gomersall, C. D. (2005). Protecting healthcare staff from severe acute respiratory syndrome: filtration capacity of multiple surgical masks. *Journal of Hospital Infection*. 59: 365-368.

Dighe, A., Jombart, T., Van Kerkhove, M. D., & Ferguson, N. (2019). A systematic review of MERS-CoV seroprevalence and RNA prevalence in dromedary camels: Implications for animal vaccination. *Epidemics*. 29: 100350.

Ding, Y. Q., He, L., Zhang, Q., Huang, Z., Che, X., Hou, J. L., Wang, H., Shen, H., Qiu, L., Li, Z., Geng, J., Cai, J., Han, H., Li, X., Kang, W., Weng, D., Liang, P., & Jiang, S. (2004). Organ distribution of severe acute respiratory syndrome (SARS) associated coronavirus (SARS-CoV) in SARS patients: implications for pathogenesis and virus transmission pathways. *The Journal of Pathology*. 203(2): 622-630.

Dolan, S. A. (2003). A new disease emerges: severe acute respiratory syndrome (SARS). *Journal for Specialists in Pediatric Nursing*. 8(2): 75-76.

Douglas, M. G., Kocher, J. F., Scobey, T., Baric, R. S., & Cockrell, A. S. (2018). Adaptive evolution influences the infectious dose of MERS-CoV necessary to achieve severe respiratory disease. *Virology*. 517: 98-107.

Du, L. Y., He, Y. X., Zhou, Y. S., Liu, S. W., Zheng, B. J., & Jiang, S. B. (2009). The spike protein of SARS-CoV- a target for vaccine and therapeutic development. *Nat. Rev. Microbiol*. 7: 226-236.

Du, L., Kou, Z., Ma, C., Tao, X., Wang, L., and Zhao, G., et al. (2013). A truncated receptor-binding domain of MERS-CoV spike protein potently inhibits MERS-CoV infection and induces strong neutralizing antibody responses: implication for developing therapeutics and vaccines. *PLOS ONE*. 8: e81587.

Du, L., Yang, Y., Zhou, Y., Lu, L., Li, F., & Jiang, S. (2017). MERS-CoV spike protein: a key target for antivirals. *Expert Opin Ther Targets*. 21: 131-143.

Du, H. Z., Hou, X. Y., Miao, Y. H., Huang, D. S., & Liu, D. H. (2019). Traditional Chinese medicine: an effective treatment for 2019 novel coronavirus pneumonia (NCP). *Chinese Journal of Natural Medicines*. 18(3): 206-210.

Dyall, J., Cross, R., Kindrachuk, J., Johnson, R. F., Olinger, G. G., Hensley, L. E., Frieman, M. B., & Jahrling, P. B. (2017). Middle East respiratory syndrome and severe acute respiratory syndrome: current therapeutic options and potential targets for novel therapies. *Drugs*. 77(18): 1935-1966.

Ebihara, H., Groseth, A., Neumann, G., Kawaoka, Y., & Feldmann, H. (2005). The role of reverse genetics system in studying viral hemorrhagic fevers. *Thromb. Haemost*. 94(2): 240-253.

Eifan, S. A., Nour, I., Hanif, A., Zamzam, A. M. M., & Aljohani, S. M. (2017). A pandemic risk assessment of Middle East respiratory syndrome coronavirus (MERS-CoV) in Saudi Arabia. *Saudi Journal of Biological Sciences*. 24: 1631-1638.

El-Zowalaty, M. E., & Jarhult, J. D. (2020). From SARS to COVID-19: a previously unknown SARS-related coronavirus (SARS-CoV-2) of pandemic potential infecting humans- Call for a One Health approach. *One Health*. 9: 100124.

Enjuanes, L., Sune, F., Gebauer, C., Smerdou, C., Camacho, A., Anton, I. M., Gonzalez, S., Talamillo, A., Mendez, A., & Ballesteros, M. L., et al., (1992). Antigen selection and presentation to protect against transmissible gastroenteritis coronavirus. *Vet. Microbiol*. 33: 249.

Faridi, U. (2018). Middle East respiratory syndrome coronavirus (MERS-CoV): impact on Saudi Arabia, 2015. *Saudi Journal of Biological Sciences*. 25: 1402-1405.

Fujii, T., Nakamura, T., & Iwamoto, A. (2004). Current concepts in SARS treatment. *J Infect Chemother*. 10: 1-7.

Gallagher, T. M., & Buchmeier, M. J. (2001). Coronavirus spike proteins in viral entry and pathogenesis. *Virology*. 279: 371.

Gebauer, F., Posthumus, W. P., Correa, I., Sune, C., Smerdou, C., Sanchez, C. M., Lenstra, J. A., Meloen, R. H., & Enjuanes, L. (1991). Residues involved in the antigenic sites of transmissible gastroenteritis coronavirus S glycoprotein. *Virology*. 183: 225.

Ghinai, I., McPherson, T. D., Hunter, J. C., Kirking, H. L., Christiansen, D., Joshi, K., Rubin, R., Morales-Estrada, S., Black, S. R., Pacilli, M., Fricchione, M. J., Chugh, R. K., Walblay, K. A., Ahmed, N. S., Stoecker, W. C., Hasan, N. F., Burdsall, D. P., Reese, H. E., Wallace, M., Wang, C., Moeller, D., Korpics, J., Novosad, S. A., Benowitz, I., Jacobs, M. W., Dasari, V. S., Patel, M. T., Kauerauf, J., Charles, E. M., Ezike, N. O., Chu, V., Midgley, C. M., Rolfes, M. A., Gerber, S. I., Lu, X., Lindstrom, S., Verani, J. R., & Layden, J. E. (2020). First known person-to-person transmission of severe acute respiratory syndrome coronavirus 2 (SARS-CoV-2) in the USA. *Lancet*. http://doi.org/10.1016/S0140-6736(20)30607-3

Gillissen, A., & Ruf, B. R. (2003). Severe acute respiratory syndrome (SARS). *Medizinische Klinik*. 98(6): 319-325.

Gralinski, L. E., & Menachery, V. D. (2020). Return of the Coronavirus: 2019-nCoV. *Viruses*. 12: 135. Doi: 10.3390/v12020135

Greenberg, S. B. (2016). Update on human rhinovirus and coronavirus infections. *Semin. Respir. Crit. Care Med*. 37: 555-571.

Grubaugh, N. D., Petrone, M. E., & Holmes, E. C. (2020). We should not worry when a virus mutates during disease outbreaks. *Nat Microbiol.* Published online February 18, 2020. http://doi.org/10.1038/s41564-020-0690-4

Gu, J., & Korteweg, C. (2007). Pathology and pathogenesis of severe acute respiratory syndrome. *The American Journal of Pathology.* 170(4): 1136-1147.

Guo, X., Deng, Y., Chen, H., Lan, J., Wang, W., & Zou, X., et al., (2015). Systemic and mucosal immunity in mice elicited by a single immunization with human adenovirus type 5 or 41 vector-based vaccines carrying the spike protein of Middle East respiratory syndrome coronavirus. *Immunology.* 145: 476-484.

Guo, H., Zhou, Y., Liu, X., & Tan, J. (2020). The impact of the COVID-19 epidemic on the utilization of emergency dental services. *Journal of Dental Sciences.* http://doi.org/10.1016/j.jds.2020.02.002.

Hafeez, R., Aslam, M., Aman, S., & Tahir, M. (2016). Severe acute respiratory syndrome (SARS): a deadly disease. *Annals of King Edward Medical University Lahore Pakistan.* 10(1).

Han, Q., Lin, Q., Jin, S., & You, L. (2020). Coronavirus 2019-nCoV: A brief perspective from the front line. *Journal of Infection.* 80: 373-377.

Harrison, S. C. (2008). Viral membrane fusion. *Nat. Struct. Mol. Biol.* 15: 690-698.

Hashem, A. M., Al-Amri, S. S., Al-Subhi, T. L., Siddiq, L. A., Hassan, A. M., Alawi, M. M. Alhabbab, R. Y., Hindawi, S. I., Mohammed, O. B., Amor, N. S., Alagaili, A. N., Morza, A. A., & Azhar, E. I. (2019). Development and validation of different indirect ELISAs for MERS-CoV serological testing. *Journal of Immunological Methods.* 466: 41-46.

Haverkamp, A. K., Bosch, B. J., Spitzbarth, I., Lehmbecker, A., Te, N., Bensaid, A., Segales, J., & Baumgartner, W. (2019). Detection of MERS-CoV antigen on formalin-fixed paraffin-embedded nasal tissue of alpacas by immunohistochemistry using human monoclonal antibodies directed against different epitopes of the spike protein. *Veterinary Immunology and Immunopathology.* 218: 109939.

Hawkey, P. M., Bhagani, S., & Gillespie, S. H. (2003). Severe acute respiratory syndrome (SARS): breath-taking progress. *Journal of Medical Microbiology*. 52: 609-613.

He, Y., Zhou, Y., Wu, H., Luo, B., Chen, J., Li, W., & Jiang, S. (2004). Identification of immunodominant sites on the spike protein of severe acute respiratory syndrome (SARS) coronavirus: implication for developing SARS diagnostics and vaccines. *The Journal of Immunology*. 173: 4050-4057.

Hijawi, B., Abdallat, M., Sayaydeh, A., Alqasrawi, S., Haddadin, A., & Jaarour, N, et al., (2013). Novel coronavirus infections in Jordan, April 2012: epidemiological findings from a retrospective investigation. *East Mediterr Heal J*. 19(Suppl 1): S12-8.

Hoheisel, G., Luk, W. K., Winkler, J., Gillissen, A., Wirtz, H., Liebert, Y., & Hui, D. S. (2007). Severe acute respiratory syndrome (SARS). *Medizinische Klinik*. 101(12): 957-963.

Holmes, K. V. (2003). SARS coronavirus: a new challenge for prevention and therapy. *J. Clin. Invest*. 111: 1605.

Holmes, K. V. (2003). SARS-associated coronavirus. *N. Engl. J. Med*. 348: 1948.

Holshue, M. L., DeBolt, C., Lindquist, S., Lofy, K. H., Wiesman, J., & Bruce, H., et al., (2020). First case of 2019 novel coronavirus in the United States. *N Engl J Med*.

Hong, X., Currier, G. W., Zhao, X., Jiang, Y., Zhou, W., & Wei, J. (2009). Posttraumatic stress disorder in convalescent severe acute respiratory syndrome patients: a 4-year follow-up study. *General Hospital Psychiatry*. 31: 546-554.

Hsih, W. H., Cheng, M. Y., Ho, M. W., Chou, C. H., Lin, P. C., Chi, C. Y., Liao, W. C., Chen, C. Y., Leong, L. Y., Tien, N., Lai, H. C., Lai, Y. C., & Lu, M. C. (2020). Featuring COVID-19 cases via screening symptomatic patients with epidemiologic link during flu season in a medical center of central Taiwan. *Journal of Microbiology, Immunology and Infection*. http://doi.org/10.1016/j.jmii.2020.03.008

Huang, J., Cao, Y., Du, J., Bu, X., Ma, R., & Wu, C. (2007). Printing with SARS CoV S DNA and boosting with SARS CoV S epitopes specific for $CD4^+$ and $CD8^+$ T cells promote cellular immune responses. *Vaccine*. 25: 6981-6991.

Hui, D. S. C., & Chan, P. K. S. (2010). Severe acute respiratory syndrome and coronavirus. *Infectious Disease Clinics of North America*. 24(3): 619-638.

Hui, D. S., Perlman, S., & Zumla, A. (2015). Spread of MERS to South Korea and China. *Lancet Respir Med*. 3: 509-510.

Hui, D. S., Azhar, E., Kim, Y. J., Memish, Z. A., Oh, M. D., & Zumla, A. (2018). Middle East respiratory syndrome coronavirus: risj factors and determinants of primary, household, and nosocomial transmission. *Lancet Infect Dis*. 18: e217-227.

Hui, D. S. C., & Zumla, A. (2019). Severe acute respiratory syndrome: historical, epidemiologic, and clinical features. *Infectious Disease Clinics of North America*. 33(4): 869-889.

Inn, K. S., Kim, Y., Aigerim, A., Park, U., Hwang, E. S., Choi, M. S., Kim, Y. S., & Cho, N. H. (2018). Reduction of soluble dipeptidyl peptidase 4 levels in plasma of patients infected with Middle East respiratory syndrome coronavirus. *Virology*. 518: 324-327.

Ji, W., Wang, W., Zhao, X., Zai, J., & Li, X. (2020). Cross species transmission of the newly identified coronavirus 2019-nCoV-Ji-2020. Journal of Medical Virology, Wiley Online Library. *J Med Virol*. 92: 433-440.

Jiang, X., Deng, L., Zhu, Y., Ji, H., Tao, L., Liu, L., Yang, D., & Ji, W. (2020). Psychological crisis intervention during the outbreak period of new coronavirus pneumonia from experience in Shanghai. *Psychiatry Research*. 286: 112903.

Jo, S., Kim, H., Kim, S., Shin, D. H., & Kim, M. S. (2019). Characteristics of flavonoids as potent MERS-CoV 3c-like protease inhibitors. *Chem Biol Drug Des*. 94: 2023-2030.

Jung, S. Y., Kang, K. W., Lee, E. Y., Seo, D. W., Kim, H. L., Kim, H., Kwon, T. W., Park, H. L., Kim, H., Lee, S. M., & Nam, J. H. (2018). Heterologous prime-boost vaccination with adenoviral vector and protein nanoparticles induces both Th1 and Th2 responses against

Middle East respiratory syndrome coronavirus. *Vaccine*. 36: 3468-3476.

Kam, Y. W., Kien, F., Roberts, A., Cheung, Y. C., Lamirande, E. W., Vogel, L., Chu, S. L., Tse, J., Guarner, J., Zaki, S. R., Subbarao, K., Peiris, M., Nal, B., & Altmeyer, R. (2007). Antibodies against trimeric S glycoprotein protect hamsters against SARS-CoV challenge despite their capacity to mediate FcγRII-dependent entry into B cells *in vitro*. *Vaccine*. 25: 729-740.

Kang, S., Peng, W., Zhu, Y., Lu, S., Zhou, M., Lin, W., Wu, W., Huang, S., Jiang, L., Luo, X., & Deng, M. (2020). Recent progress in understanding 2019 novel coronavirus associated with human respiratory disease: detection, mechanism and treatment. *International Journal of Antimicrobial Agents*. Doi: http://doi.org/10.1016/j.ijantimicag.2020.105950

Karypidou, K., Ribone, S. R., Quevedo, M. A., Persoons, L., Pannecouque, C., Helsen, C., Claessens, F., & Dehaen, W. (2018). Synthesis, biological evaluation and molecular modeling of a novel series of fused 1,2,3-triazoles as potential anti-coronavirus agents. *Bioorganic and Medicinal Chemistry Letters*. 28: 3472-3476.

Kasem, S., Qasim, I., Al-Hufofi, A., Hashim, O., Alkarar, A., Abu-Obeida, A., Gaafer, A., Elfadil, A., Zaki, A., Al-Romaihi, A., Babekr, N., El-Harby, N., Hussien, R., Al-Sahaf, A., Al-Doweriej, A., Bayoumi, F., Poon, L. L. M., Chu, D. K. W., Peiris, M., & Perera, R. A. P. M. (2018). Cross-sectional study of MERS-CoV-specific RNA and antibodies in animals that have had contact with MERS patients in Saudi Arabia. *Journal of Infection and Public Health*. 11: 331-338.

Keyaerts, E., Vijgen, L., Maes, P., Neyts, J., & Ranst, M. V. (2004). *In vitro* inhibition of severe acute respiratory syndrome coronavirus by chloroquine. *Biochemical and Biophysical Research Communications*. 323: 264-268.

Khan, M. U., Shah, S., Ahmad, A., & Fatokun, O. (2014). Knowledge and attitude of healthcare workers about Middle East Respiratory Syndrome in multispecialty hospitals of Qassim, Saudi Arabia. *BMC Public Health*. 14: 1281.

Khan, S., Nabi, G., Han, G., Siddique, R., Lian, S., Shi, H., Bashir, N., Ali, A., & Adnan Shereen, M. (2020). Novel coronavirus: how things are in Wuhan. *Clinical Microbiology and Infection.* http://doi.org/10.1016/j.cmi.2020.02.005.

Kharma, M. Y., Alalwani, M. S., Amer, M. F., Tarakji, B., & Aws, G. (2015). Assessing of the awareness level of dental students toward Middle East Respiratory Syndrome-coronavirus. *J Int Soc Prev Community Dent.* 5(3): 163-169.

Kim, E., Okada, K., Kenniston, T., Raj, V. S., AlHajri, M. N., & Farag, E. A., et al., (2014). Immunogenicity of an adenoviral-based Middle East respiratory syndrome coronavirus vaccine in BALB/c mice. *Vaccine.* 32: 5975-5982.

Kim, Y., Lee, S., Chu, C., Choe, S., Hong, S., & Shin, Y. (2016). The characteristics of Middle Eastern respiratory syndrome coronavirus transmission dynamics in South Korea. *Osong Public Heal Res Perspect.* 7: 49-55.

Kim, Y. (2018). Nurses experiences of care for patients with Middle East respiratory syndrome-coronavirus in South Korea. *American Journal of Infection Control* 46: 781-787.

Kissoon, N. (2003). The threat of severe acute respiratory syndrome (SARS). *The West Indian Medical Journal.* 52(2): 91-94.

Ko, J. H., Kim, S. H., Lee, N. Y., Kim, Y. J., Cho, S. Y., Kang, C. I., Chung, D. R., & Peck, K. R. (2019). Effects of environmental disinfection on the isolation of vancomycin-resistant *Enterococcus* after a hospital-associated outbreak of Middle East respiratory syndrome. *American Journal of Infection Control.* 47: 1516-1518.

Kobayashi, T., Jung, S. M., Linton, N. M., Kinoshita, R., Hayashi, K., Miyama, T., Anzai, A., Yang, Y., Yuan, B., Akhmetzhanov, A. R., Suzuki, A., & Nishiura, H. (2020). Communicating the risk of death from novel coronavirus disease (COVID-19). *Journal of Clinical Medicine.* 9: 580.

Kucharski, A. J., & Althaus, C. L. (2015). The role of superspreading in Middle East respiratory syndrome coronavirus (MERS-CoV) transmission. *Eurosurveillance.* 20(25).

Kumaki, Y., Wandersee, M. K., Smith, A. J., Zhou, Y., Simmons, G., Nelson, N. M., Bailey, K. W., Vest, Z. G., Li, J. K. K., Chan, P. K. S., Smee, D. F., & Barnard, D. L. (2011). Inhibition of severe acute respiratory syndrome coronavirus replication in a lethal SARS-CoV BALB/c mouse model by stinging nettle lectin, *Urtica dioica* agglutinin. *Antiviral Research*. 90: 22-32.

Kuo, L., & Masters, P. S. (2003). The small envelope protein E is not essential for murine coronavirus replication. *J. Virol*. 77: 4597-4608.

Lagier, J. C., Colson, P., Dupont, H. T., Salomon, J., Doudier, B., Aubry, C., Gouriet, F., Baron, S., Dudouet, P., Flores, R., Ailhaud, L., Gautret, P., Parola, P., La Scola, B., Raoult, D., & Brouqui, P. (2020). Testing the repatriated SARS-CoV2: should laboratory-based quarantine replace traditional quarantine? *Travel Medicine and Infectious Disease*. http://doi.org/10.1016/j.tmaid.2020.101624.

Lai, C. C., Liu, Y. H., Wang, C. Y., Wang, Y. H., Hsueh, S. C., Yen, M. Y., Ko, W. C., & Hsueh, P. R. (2020). Asymptomatic carrier state, acute respiratory disease, and pneumonia due to severe acute respiratory syndrome coronaviurs 2(SARS-CoV-2): fact and myths. *Journal of Microbiology, Immunology and Infection*. http://doi.org/ 10.1016/j.jmii.2020.02.012.

Lang, Z. W., Zhang, L. J., Zhang, S. J., Meng, X., Li, J. Q., Song, C. Z., Sun, L., Zhou, Y. S., & Dwyer, D. E. (2003). A clinicopathological study of three cases of severe acute respiratory syndrome (SARS). *Pathology*. 35(6): 526-531.

Lau, E. H. Y., Cowling, B. J., Muller, M. P., Ho, L. M., Tsang, T., Lo, S. V., Louie, M., & Leung, G. M. (2009). Effectiveness of ribavirin and corticosteroids for severe acute respiratory syndrome. *The American Journal of Medicine*. 122(12).

Law, P. Y. P., Liu, Y. M., Geng, H., Kwan, K. H., Waye, M. M. Y., & Ho, Y. Y. (2006). Expression and functional characterization of the putative protein 8b of the severe acute respiratory syndrome-associated coronavirus. *FEBS Letters*. 580: 3643-3648.

Lee, A., & Cho, J. (2017). The impact of city epidemics on rural labor market: The Korean Middle East Respiratory Syndrome case. *Japan and the World Economy*. 43: 30-40.

Lee, J. Y., Kim, G., Lim, D. G., Jee, H. G., Jang, Y., Joh, J. S., Jeong, I., Kim, Y., Kim, E., & Chin, B. S. (2019). A Middle East respiratory syndrome screening clinic for health care personnel during the 2015 Middle East respiratory syndrome outbreak in South Korea: A single-center experience. *American Journal of Infection Control*. 46: 436-440.

Letko, M., Miazgowicz, K., McMinn, R., Seifert, S. N., Sola, I., Enjuanes, L., Carmody, A., van Doremalen, N., & Munster. V. (2018). Adaptive evolution of MERS-CoV to species variation in DPP4. *Cell Reports*. 24: 1730-1737.

Leung, W. K., To, K. F., Chan, P. K. S., Chan, H. L. Y., Wu, A. K. L., Lee, N., Yuen, K. Y., & Sung, J. J. Y. (2003). Enteric involvement of severe acute respiratory syndrome associated coronavirus infection. *Gastroenterology*. 125: 1011-1017.

Li, G., Chen, X., & Xu, A. (2003). Profile of specific antibodies to the SARS-associated coronavirus. *N. Engl. J. Med*. 349: 508.

Li, G. M., Li, Y. G., Yamate, M., Li, S. M., & Ikuta, K. (2007). Lipid rafts play an important role in the early stage of severe acute respiratory syndrome-coronavirus life cycle. *Microbes and Infection*. 9: 96-102.

Li, W., Hulswit, R. J. G., Widjaja, I., Raj, V. S., McBride, R., Peng, W., Widagdo, W., Tortotici, M. A., van Dieren, B., Lang, Y., van Lent, J. W. M., Paulson, J. C., de Haan, C. A. M., de Groot, R. J., van Kuppeveld, F. J. M., Haagmans, B. L., & Bosch, B. J. (2017). Identification of sialic acid-binding function for the Middle East respiratory syndrome coronavirus spike glycoprotein. *Proc. Natl. Acad. Sci. USA*. 114: E8508-E8517.

Li, T. (2020). Diagnosis and clinical management of severe acute respiratory syndrome Coronavirus 2 (SARS-CoV-2) infection: an operational recommendation of Peking Union Medical College Hospital (V2.0). *Emerging Microbes and Infections*. 9(1): 582-585.

Li, L., Yang, Z., Dang, Z., Meng, C., Huang, J., Meng, H., Wang, D., Chen, G., Zhang, J., Peng, H., & Shao, Y. (2020). Propagation analysis and prediction of the COVID-19. *Infectious Disease Modelling*. 5: 282-292.

Li, C., Yang, Y., & Ren, L. (2020). Genetic evolution analysis of 2019 novel coronavirus and coronavirus from other species. *Infection, Genetics and Evolution*. 82: 104285.

Lien, T. C., Sung, C. S., Lee, C. H., Kao, H. K., Huang, Y. C., Liu, C. Y., Perng, R. P., & Wang, J. H. (2008). Characteristic features and outcomes of severe acute respiratory syndrome found in severe acute respiratory syndrome intensive care unit patients. *Journal of Critical Care*. 23: 557-564.

Lin, Q., Zhao, S., Gao, D., Lou, Y., Yang, S., Musa, S. S., Wang, M. H., Cai, Y., Wang, W., Yang, L., & He, D. (2020). A conceptual model for the coronavirus disease 2019 (COVID-19) outbreak in Wuhan, China with individual reaction and governmental action. *International Journal of Infectious Disease*. 93: 211-216.

Lippi, G., & Plebani, M. (2020). Procalcitonin in patients with severe coronavirus disease 2019 (COVID-19): A meta-analysis. *Clinica Chimica Acta*. 505: 190-191.

Lippi, G., Plebani, M., and Henry, B. M. (2020). Thrombocytopenia is associated with severe coronavirus disease 2019 (COVID-19) infections: A meta-analysis. *Clinica Chimica Acta*. 506: 145-148.

Liu, D. X., Fung, T. S., Chong, K. K. L., Shukla, A., & Hilgenfeld, R. (2014). Accessory proteins of SARS-CoV and other coronaviruses. *Antivir. Res*. 109: 97-109.

Liu, K. C., Xu, P., Lv, W. F., Qiu, X. H., Yao, J. L., Gu, J. F., & Wei, W. (2020). CT manifestations of coronavirus disease-2019: A retrospective analysis of 73 cases by disease severity. *European Journal of Radiology*. 126: 108941.

Liu, H., Wang, L. L., Zhao, S. J., Kwak-Kim, J., Mor, G., & Liao, A. H. (2020). Why are pregnant women susceptible to COVID-19? An immunological viewpoint. *Journal of Reproductive Immunology*. 139: 103122.

Long, C., Xu, H., Shen, Q., Zhang, X., Fan, B., Wang, C., Zeng, B., Li, Z., Li, X., & Li, H. (2020). Diagnosis of the coronavirus disease (COVID-19): rRT-PCR or CT? *European Journal of Radiology*. 126: 108961.

Lu, J. H., Zhang, D. M., Wang, G. L., Guo, Z. M., Zhang, C. H., Tan, B. Y., Ouyang, L. P., Lin, L., Liu, Y. M., Chen, W. Q., Ling, W. H., Yu, X. B., & Zhong, N. S. (2005). Variation analysis of the severe acute respiratory syndrome coronavirus putative non-structural protein 2 gene and construction of three-dimensional model. *Chinese Medical Journal*. 118(9): 707-713.

Lu, G., Hu, Y., Wang, Q., Qi, J., Gao, F., Li, Y., Zhang, Y., Zhang, W., Yuan, Y., Bao, J., Zhang, B., Shi, Y., Yan, J., & Gao, G. F. (2013). Molecular basis of binding between novel human coronavirus MERS-CoV and its receptor CD26. *Nature*. 500: 227-231.

Lu, G., Wang, Q., & Gao, G. F. (2015). Bat-to-human: spike features determining host jump of coronaviruses SARS-CoV, MERS-CoV, and beyond. *Trend in Microbiology*. 23(8): 468-478.

Lu, D., Wang, H., Yu, R., Yang, H., & Zhao, Y. (2020). Integrated infection control strategy to minimize nosocomial infection of coronavirus disease 2019 among ENT healthcare workers. *Journal of Hospital Infection*. http://doi.org/10.1016/j.jhin.2020.02.018.

Lu, R., Zhao, X., Li, J., Niu, P., Yang, B., Wu, H., Wang, W., Song, H., Huang, B., Zhu, N., Bi, Y., Ma, X., Zhan, F., Wang, L., Hu, T., Zhou, H., Hu, Z., Zhou, W., Zhao, W., Zhao, L., Chen, J., Meng, Y., Wang, J., Lin, Y., Yuan, J., Xie, Z., Ma, J., Liu, W., Wang, D., Xu, W., Holmes, E. C., Gao, G. F., Wu, G., Chen, W., Shi, W., & Tan, W. (2020). Genomic characterization and epidemiology of 2019 novel coronavirus: implications for the virus origins and receptor binding. *Lancet*. 395: 565-574.

Luan, J., Lu, Y., Jin, X., & Zhang, L. (2020). Spike protein recognition of mammalian ACE2 predicts the host range and an optimized ACE2 for SARS-CoV-2 infection. *Biochemical and Biophysical Research Communications*. http://doi.org/10.1016/j.bbrc.2020.03.047

Lupia, T., Scabini, S., Pinna, S. M., Di Perri, G., De Rosa, F. G., & Corcione, S. (2020). 2019 novel coronavirus (2019-nCoV) outbreak: A new challenge. *Journal of Global Antimicrobial Resistance*. 21: 22-27.

Lv, D. F., Ying, Q. M., Weng, Y. S., Shen, C. B., Chu, J. G., Kong, J. P., Sun, D. H., Gao, X., Weng, X. B., & Chen, X. Q. (2020). Dynamic change process of target genes by RT-PCR testing of SARS-CoV-2 during the course of a Coronavirus Disease 2019 patient. *Clinica Chimica Acta*. 506: 172-175.

Ma, C., Li, Y., Wang, L., Zhao, G., Tao, X., & Tseng, C. T., et al., (2014). Intranasal vaccination with recombinant receptor-binding domain of MERS-CoV spike protein induces much stronger local mucosal immune responses than subcutaneous immunization: implication for designing novel mucosal MERS vaccines. *Vaccine*. 32(18): 2100-2108.

Mailles, A., Blanckaert, K., Chaud, P., van der Werf, S., Lina, B., & Caro, V., et al., (2013). First cases of Middle East Respiratory Syndrome Coronavirus (MERS-CoV) infections in France, investigations and implications for the prevention of human-to-human transmission, France, May 2013. *Euro Surveill*. 18(24).

Masters, P. S., & Perlman, S. (2013). *Coronaviridae in Field's virology*, in: D. M. Knipe, P. M. Howley (Eds.), Lippincott, vol. 1, Williams & Wilkins, Philadelphia, pp. 825-858.

Masters, P. S. (2019). Coronavirus genomic RNA packaging. *Virology*. 537: 198-207.

Matsuyama, S., Nao, N., Shirato, K., Kawase, M., Saito, S., Takayama, I., Nagata, N., Sekizuka, T., Katoh, H., Kato, F., Sakata, M., Tahara, M., Kutsuna, S., Ohmagari, N., Kuroda, M., Suzuki, T., Kageyama, T., & Takeda, M. (2020). Enhanced isolation of SARS-CoV-2 by TMPRSS2-expressing cells. 2020. *Proc Natl Acad Sci USA*. doi: 1073/pnas.2002589117.

Maxwell, C., McGeer, A., Tai, K. F. Y., & Sermer, M. (2017). N. 225-management guidelines for obstetric patients and neonates born to mothers with suspected or probable severe acute respiratory syndrome (SARS). *J Obstet Gynaecol Can*. 39(8): e130-e137.

Memish, Z. A., Zumla, A. I., & Assiri, A. (2013). Middle East respiratory syndrome coronavirus infections in health care workers. *N Eng J Med.* 369: 884-886.

Memish, Z. A., & Al-Tawfiq, J. A. (2014). Middle East respiratory syndrome coronavirus infection control: the missing piece? *Am J Infect Control*. 42.

Memish, Z. A., Cotton, M., Watson, S. J., Kellam, P., Zumla, A., & Alhakeem, R. F., et al., (2014). Communisty case cluster of Middle East respiratory syndrome coronavirus in Hafr Al-Batin. Kingdom of Saudi Arabia: a descriptive genomic study. *Int J Infect Dis*. 23: 63-68.

Memish, Z. A., Perlman, S., van Kerkove, M. D., & Zumla, A. (2020). Middle East respiratory syndrome. *Lancet.* 395: 1063-1077.

Meyer, B., Muller, M. A., Corman, V. M., Reusken, C. B., Ritz, D., Godeke, G. J., Lattwein, E., Kallies, S., Siemens, A., van Beek, J., Drexler, J. F., Muth, D., Bosch, B. J., Wernery, U., Koopmans, M. P., Wernery, R., & Drosten, C. (2014). Antibodies against MERS coronavirus in dromedary camels, United Arab Emirates, 2003 and 2013. *Emerg Infect Dis*. 20(4): 552-559.

Meyerholz, D. K., Lambertz, A. M., & McCray, Jr. P. B. (2016). Dipeptidyl peptidase 4 distribution in the human respiratory tract. *Am J Pathol.* 186(1): 78-86.

Millet, J. K., & Whittaker, G. R. (2018). Physiological and molecular triggers for SARS-CoV membrane fusion and entry into host cells. *Virology*. 517: 3-8.

Momattin, H., Al-Ali, A. Y., & Al-Tawfiq, J. A. (2019). A systematic review of therapeutic agents for the treatment of the Middle East respiratory syndrome coronavirus (MERS-CoV). *Travel Med Infect* Dis. 30: 9-18.

Muller, M. A., Meyer, B., Corman, V. M., Al-Masri, M., Turkestani, A., Ritz, D., Sieberg, A., Aldabbagh, S., Bosch, B. J., Lattwein, E., Alhakeem, R. F., Assiri, A. M., Albarrak, A. M., Al-Shangiti, A. M., Al-Tawfiq, J. A., Wikramaranta, P., Alrabeeah, A. A., Drosten, C., & Memish, Z. A. (2015). Presence of Middle East respiratory syndrome coronavirus antibodies in Saudi Arabia: a nationwide, cross-sectional, serological study. *Lancet Infect Dis*. 15(5): 559-564.

Munster, V. J., Adney, D. R., van Doremalen, N., Brown, V. R., Miazgowicz, K. L., Milne-Price, S., Bushmaker, T., Rosenke, R., Scott, D., Hawkinson, A., de Wit, E., Schountz, T., & Bowen, R. A. (2016). Replication and shedding of MERS-CoV in Jamaican fruit bats (*Artibeus jamaicensis*). *Sci. Rep.* 6: 21878.

Mustafa, S., Balkhy, H., & Gabere, M. N. (2018). Current treatment options and the role of peptides as potential therapeutic components for Middle East Respiratory Syndrome (MERS): A review. *Journal of Infection and Public Health*. 11: 9-17.

Muthumani, K., Falzarano, D., Reuschel, E. L., Tingey, C., Flingai, S., & Villarreal, D. O, et al., (2015). A synthetic consensus anti-spike protein DNA vaccine induces protective immunity against Middle East respiratory syndrome coronavirus in nonhuman primates. *Sci Transl Med*. 7(301): 301ra132.

Nah, K., Otsuki, S., Chowell, G., & Nishiura, H. (2016). Predicting the international spread of Middle East respiratory syndrome (MERS). *BMC Infectious Diseases*. 16: 356.

Narayanan, K., Huang, C., & Makino, S. (2008). SARS coronavirus accessory proteins. *Virus Research*. 133: 113-121.

Naz, R. K., & Dabir, P. (2006). Peptide vaccines against cancer, infectious diseases, and conception. *Front Biosci*. 12: 1833-1844.

Netland, J., DeDiego, M. L., Zhao, J., Fett, C., Alvarez, E., Nieto-Torres, J. L., Enjuanes, L., & Perlman, S. (2010). Immunization with an attenuated severe acute respiratory syndrome coronavirus deleted in E protein protects against lethal respiratory disease. *Virology*. 399: 120-128.

Nguyen, T. T. H., Ryu, H. J., Lee, S. H., Hwang, S., Breton, V., Rhee, J. H., & Kim, D. (2011). Virtual screening identification of novel severe acute respiratory syndrome 3C-like protease inhibitors and *in vitro* confirmation. *Bioorganic and Medicinal Chemistry Letters*. 21: 3088-3091.

Nie, Q. H., Luo, X. D., & Hui, W. L. (2003). Advances in clinical diagnosis and treatment of severe acute respiratory syndrome. *World J Gastroenterol.* 9(6): 1139-1143.

Nieto-Torres, J. L., DeDiego, M. L., Verdia-Baguena, C., Jimenez-Guardeno, J. M., Regla-Nava, J. A., Fernandez-Delgado, R., Castano-Rodriguez, C., Alcaraz, A., Torres, J., Aguilella, V. M., & Enjuanes, L. (2014). Severe acute respiratory syndrome coronavirus envelope protein ion channel activity promotes virus fitness and pathogenesis. *PLOS Pathog*. http://dx.doi.org/10.1371/journal.ppat.1004077.

Nikiforuk, A. M., Leung, A., Cook, B. W. M., Court, D. A., Kobasa, D., & Theriault, S. S. (2016). Rapid one-step construction of a Middle East Respiratory Syndrome (MERS-CoV) infectious clone system by homologous recombination. *Journal of Virology Methods*. 236: 178-183.

Noorwali, A. A., Turkistani, A. M., Asiri, S. I., Trabulsi, F. A., Alwafi, O. M., & Alzahrani, S. H., et al. (2015). Descriptive epidemiology and characteristics of confirmed cases of Middle East respiratory syndrome coronavirus infection in the Makkah Region of Saudi Arabia. March to June 2014. *Ann Saudi Med.* 35: 203-209.

Ohnishi, K., Hattori, Y., Kobayashi, K., & Akaji, K. (2019). Evaluation of a non-prime site substituent and warheads combined with a decahydroisoquinolin scaffold as a SARS 3CL protease inhibitor. *Bioorganic and Medicinal Chemistry*. 27: 425-435.

Ortego, J., Ceriani, J. E., Patino, C., Plana, J., & Enjuanes, L. (2007). Absence of E protein arrests transmissible gastroenteritis coronavirus maturation in the secretory pathway. *Virology*. 368(2): 296-308.

Pallesen, J., Wang, N., Corbett, K. S., Warpp, D., Kirchdoerfer, R. N., turner, H. L., Cottrell, C. A., Becker, M. M., Wang, L., Shi, W., Kong, W. P., Andres, E. L., Kettenbach, A. N., Denison, M. R., Chappell, J. D., Graham, B. S., Ward, A. B., & McLellan, J. S. (2017). Immunogenicity and structures of a rationally designed prefusion MERS-CoV spike antigen. *Proc. Natl. Acad. Sci. USA*. 114: E7348-E7357.

Pang, J., Wang, M. X., Ang, I. Y. H., Tan, S. H. X., Lewis, R. F., Chen, J. I. P., Gutierrez, R. A., Gwee, S. X. W., Chua, P. E. Y., Yang, Q., Ng, X. Y., Yap, R. K. S., Tan, H. Y., Teo, Y. Y., Tan, C. C., Cook, A. R., Yap, J. C. H., & Hsu, L. Y. (2020). Potential rapid diagnostics, vaccine and therapeutics for 2019 novel coronavirus (2019-nCoV): A systematic review. *Journal of Clinical Medicine*. 9: 623.

Papaneri, A. B., Johnson, R. F., Wada, J., Bollinger, L., Jahrling, P. B., & Kuhn, J. H. (2015). Middle East respiratory syndrome: obstacles and prospects for vaccine development. *Expert Rev. Vaccines*. 14: 949-962.

Parashar, U. D., & Anderson, L. J. (2004). Severe acute respiratory syndrome: review and lessons of the 2003 outbreak. *International Journal of Epidemiology*. 33: 628-634.

Park, H. Y., Lee, E. J., Ryu, Y. A., Kim, Y., Kim, H., & Lee, H., et al. (2015). Epidemiological investigation of MERS-CoV spread in a single hospital in South Korea, May to June 2015. *Eurosurveillance*. 20(June25): 21169.

Park, S. Y., Lee, J. S., Son, J. S., Ko, J. H., Peck, K. R., Jung, Y., Woo, H. J., Joo, Y. S., Eom, J. S., & Shi, H. (2019). Post-exposure prophylaxis for Middle East respiratory syndrome in healthcare workers. *Journal of Hospital Infection*. 101: 42-46.

Patrick, D. M. (2003). The race to outpace severe acute respiratory syndrome (SARS). *Canadian Medical Association Journal*. 168(10): 1265-1266.

Peiris, J. S. M. (2003). Severe acute respiratory syndrome (SARS). *Journal of Clinical Virology*. 28(3): 245-247.

Perlman, S., & Vijay, R. (2016). Middle East respiratory syndrome vaccines. Middle East respiratory syndrome vaccines. *International Journal of Infectious Diseases*. 47: 23-28.

Petrosillo, N., Viceconte, G., Ergonul, O., Ippolito, G., & Petersen, E. (2020). COVID-19, SARS and MERS: are they closely related? *Clinical Microbiology and Infection*. http://doi.org/10.1016/j.cmi.2020.03.026

Pillaiyar, T., Manickam, M., Namasivayam, V., Hayashi, Y., & Jung, S. H. (2016). An overview of severe acute respiratory syndrome-coronavirus

(SARS-CoV) 3CL protease inhibitors: peptidomimetics and small molecule chemotherapy. *Journal of Medicinal Chemistry*. 59: 6595-6628.

Poissy, J., Goffard, A., Parmentier-Decruq, E., Favory, R., Kauv, M., Kipnis, E., Mathieu, D., Guery, B., & The MERS-CoV Biology Group. (2014). Kinetics and pattern of viral excretion in biological specimens of two MERS-CoV cases. *Journal of Clinical Virology*. 61: 275-278.

Poletto, C., Gomes, M. F., Pastore y Piontti, A., Rossi, L., Bioglio, L., Chao, D. L., Longini, I. M., Halloran, M. E., Colizza, A., & Vespignani, A. (2014). Assessing the impact of travel restrictions on international spread of the 2014 West African Ebola epidemic. *Euro Surveill*. 19(42).

Poon, L. L. M., Chan, K. H., Nicholls, J. M., Zheng, B. J., Yuen, K. Y., Guan, Y., & Peiris, J. S. M. (2004). Characterization of a novel coronavirus responsible for severe acute respiratory syndrome. *International Congress Series*. 1263: 805-808.

Posthumus, W. P., Lenstra, J. A., van Nieuwstadt, A. P., Schaaper, W. M., van der Zeijst, B. A., & Meloen, R. H. (1991). Immunogenicity of peptides simulating a neutralization epitope of transmissible gastroenteritis virus. *Virology*. 182: 371.

Prem, K., Liu, Y., Russell, T. W., Kucharski, A. J., Eggo, R. M., & Davies, N., et al., (2020). The effect of control strategies to reduce social mixing on outcomes of the COVID-19 epidemic in Wuhan, China: a modeling study. *Lancet Public Health*. http://doi.org/10.1016/S2468-2667(20)30073-6

Qiu, H., Sun, S., Xiao, H., Feng, J., Guo, Y., Tai, Wang, Y., Du, L., Zhao, G., & Zhou, Y. (2016). Single-dose treatment with a humanized neutralizing antibody affords full protection of a human transgenic mouse model from lethal Middle East respiratory syndrome (MERS)-coronavirus infection. *Antiviral Research*. 132: 141-148.

Rabaan, A. A., Alahmed, S. H., Bazzi, A. M., & Alhani, H. M. (2017). A review of candidate therapies for Middle East respiratory syndrome from a molecular perspective. *J Med Microbiol*. 66: 1261-1274.

Rabenau, H. F., Kampf, G., Cinatl, J., & Doerr, H. W. (2005). Efficacy of various disinfectants against SARS coronavirus. *Journal of Hospital Infection*. 61(2): 107-111.

Raj, V. S., Mou, H., Smits, S. L., Dekkers, D. H. W., Muller, M. A., Dijkman, R., Muth, D., Demmers, J. A. A., Zaki, A., Fouchier, R. A. M., Thiel, V., Drosten, C., Rottier, P. J. M., Osterhaus, A. D. M. E., Bosch, B. J., & Haagmans, B. L. (2013). Dipeptidyl peptidase 4 is a functional receptor for the emerging human coronavirus-EMC. *Nature*. 495: 251-254.

Rasmussen, S. A., Smulian, J. C., Lednicky, J. A., Wen, T. S., & Jamieson, D. J. (2020). Coronavirus disease 2019 (COVID-19) and pregnancy: what obstetricians needs to know. *American Journal of Obstetrics and Gynecology*. doi: http://doi.org/10.1016/j,ajog.2020.02.017.

Rello, J., Tejada, S., Userovici, C., Arvaniti, K., Pugin, J., & Waterer, G. (2020). Coronavirus Disease 2019 (COVID-19): A critical care perspective beyond China. *Anaesth Crit Care Pain Med*. https://doi.org/10.1016/j.accpm.2020.03.001

Richardson, P., Griffin, I., Tucker, C., Smith, D., Oechsle, O., & Phelan, A., et al., (2020). Baricitinib as potential treatment for 2019-nCoV acute respiratory disease. *The Lancet*.

Robson, B. (2020). Computers and viral disease. Preliminary bioinformatics studies on the design of a synthetic vaccine and a preventive peptidomimetic antagonist against the SARS-CoV-2 (2019-nCoV, COVID-19) coronavirus. *Computers in Biology and Medicine*. 119: 103670.

Roda, W. C., Varughese, M. B., Han, D., & Li, M. Y. (2020). Why is it difficult to accurately predict the COVID-19 epidemic? *Infectious Disease Modelling*. 5: 271-281.

Rota, P. A., Oberste, M. S., Monroe, S. S., Nix, W. A., Campagnoli, R., Icenogle, J. P., Penaranda, S., Bankamp, B., Maher, K., & Chen, M. H., et al., (2003). Characterization of a novel coronavirus associated with severe acute respiratory syndrome. *Science*. 300: 1394-1399.

Rothan, H. A., & Byrareddy, S. N. (2020). The epidemiology and pathogenesis of coronavirus disease (COVID-19) outbreak. *Journal of Autoimmunity*. http://doi.org/10.1016/j.jaut.2020.102433.

Rubin, E. J., Baden, L. R., Morrissey, S., & Campion, E. W. (2020). Campion, Medical Journals and the 2019-nCoV Outbreak. *N Engl J Med*.

Ruiz-Contreras, A. (2003). Case report: caring for suspected severe acute respiratory syndrome (SARS) patients. *Disaster Management and Response*. 1(3): 71-75.

Saminathan, M., Chakraborty, S., Tiwari, R., Dhama, K., & Verma, A. (2014). Coronavirus infection in equines: a review. *Asian J. Anim. Vet. Adv*. 9: 164-176.

Schoeman, D., & Fielding, B. C. (2019). Coronavirus envelope protein: current knowledge. *Virol. J*. 16: 69.

Seah, I., & Agrawal, R. (2020). Can the coronavirus disease 2019 (COVID-19) affect the eyes? A review of coronaviruses and ocular implications in humans and animals. *Ocular Immunology and Inflammation*. http://doi.org/10.1080/09273948.2020.1738501

Shalhoub, S., Farahat, F., Al-Jiffri, A., Simhairi, R., Shamma, O., & Siddiqi, N., et al., (2015). IFN-α2a or IFN-β1a in combination with ribavirin to treat Middle East respiratory syndrome coronavirus pneumonia: a retrospective study. *Antimicrob Chemother*. 70: 2129-2132.

Sheahan, T. P., Sims, A. C., & Graham, R. L., et al., (2017). Broad-spectrum antiviral GS-5734 inhibits both epidemic and zoonotic coronaviruses. *Sci Transl Med*. 9: eaal3653.

Sheahan, T. P., Sims, A. C., Leist, S. R., Schafer, A., Won, J., & Brown, A. J., et al., (2020). Comparative therapeutic efficacy of remdesivir and combination lopinavir, ritonavir, and interferon beta against MERS-CoV. *Nat Commun*. 11(1): 1-14.

Shen, B., Cheng, S. K. W., Lau, K. K., Li, H. L., & Chan, E. L. Y. (2005). The effects of disease severity, use of corticosteroids and social factors on neuropsychiatric complaints in severe acute respiratory syndrome

(SARS) patients at acute and convalescent phases. *European Psychiatry*. 20: 236-242.

Shen, M., Zhou, Y., Ye, J., Al-maskri, A. A. A., Kang, Y., Zeng, S., & Cai, S. (2020). Recent advances and perspectives of nucleic acid detection for coronavirus. *Journal of Pharmaceutical Analysis*. http://doi.org/10.1016/j.jpha.2020.02.010.

Shereen, M. A., Khan, S., Kazmi, A., Bashir, N., & Siddique, R. (2020). COVID-19 infection: origin, transmission, and characteristics of human coronaviruses. *Journal of Advanced Research*. 24: 91-98.

Shin, G. C., Chung, Y. S., Kim, I. S., Cho, H. W., & Kang, C. (2006). Preparation and characterization of a novel monoclonal antibody specific to severe acute respiratory syndrome-coronavirus nucleocapsid protein. *Virus Research*. 122: 109-118.

Shin, G. C., Chung, Y. S., Kim, I. S., Cho, H. W., & Kang, C. (2007). Antigenic characterization of severe acute respiratory syndrome-coronavirus nucleocapsid protein expressed in insect cells: The effect of phosphorylation on immunoreactivity and specificity. *Virus Research*. 127: 71-80.

Sigrist, C. J. A., Bridge, A., & Le Mercier, P. (2020). A potential role for integrins in host cell entry by SARS-CoV-2. *Antiviral Research*. 177: 104759.

Sims, A. C., Burkett, S. E., Yount, B., & Pickles, R. J. (2008). SARS-CoV replication and pathogenesis in an *in vitro* model of the human conducting airway epithelium. *Virus Research*. 133: 33-44.

Snijder, E. J., Bredenbeek, P. J., Dobbe, J. C., Thiel, V., Ziebuhr, J., Poon, L. L., Guan, Y., Rozanov, M., Spaan, W. J., & Gorbalenya, A. E. (2003). Unique and conserved features of genome and proteome of SARS-coronavirus, an early split-off from the coronavirus group 2 lineage. *J. Mol. Biol*. 331: 991-1004.

Sohrabi, C., Alsafi, Z., O'Neill, N., Khan, M., Kerwan, A., Al-Jabir, A., Iosifidis, C., & Agha, R. (2020). World Health Organization declares global emergency: a review of the 2019 novel coronavirus (COVID-19). *International Journal of Surgery*. 76: 71-76.

Song, F., Fux, R., Provacia, L. B., Volz, A., Eickmann, M., & Becker, S., et al., (2013). Middle East respiratory syndrome coronavirus spike protein delivered by modified vaccine virus Ankara efficiently induces virus-neutralizing antibodies. *J Virl.* 87: 11950-11954.

Spiegel, M., Pichlmair, A., Muhlberger, E., Haller, O., & Weber, F. (2004). The antiviral effect of interferon-beta against SARS-coronavirus is not mediated by MxA protein. *J. Clin. Virol. Off. Publ. Pan Am. Soc. Clin. Virol.* 30: 211-213.

Spiegel, M., Pichlmair, A., Martinez-Sobrido, K., Cros, J., Garcia-Sastre, A., Haller, O., & Weber, F. (2005). Inhibition of Beta interferon induction by severe acute respiratory syndrome coronavirus suggests a two-step model for activation of interferon regulatory factor 3. *J. Virol.* 79: 2079-2086.

Stockman, L. J., Bellamy, R., & Garner, P. (2006). SARS: systematic review of treatment effects. *PLOS Medicine.* 3(9): e343. DOI: 10.1371/journal.pmed.0030343.

Stohr, K. (2003). World Health Organization multicentre collaborative network for severe acute respiratory syndrome (SARS) diagnosis. A multi-centre collaboration to investigate the cause of severe acute respiratory syndrome. *The Lancent.* 361: 1730-1733.

Sun, K., Chen, J., & Viboud, C. (2020). Early epidemiological analysis of the coronavirus disease 2019 outbreak based on crowdsourced data: a population-level observational study. *Lancet Digital Health.* 2: e201-208.

Sun, Z., Thilakavathy, K., Kumar, S. S., He, G., & Liu, S. V. (2020). Potential factors influencing repated SARS outbreaks in China. *International Journal of Environmental Research and Public Health.* 17: 1633.

Tai, D. Y. H. (2007). Pharmacologic treatment of SARS: current knowledge and recommendations. *Annals Academy of Medicine.* 36(6): 438-443.

Tang, J., Zhang, N., Tao, X., Zhao, G., Guo, Y., & Tseng, C. T., et al., (2015). Optimization of antigen dose for a receptor-binding domain-based subunit vaccine against MERS coronavirus. *Hum Vaccin Immunother*. 11: 1244-1250.

Tang, B., Bragazzi, N. L., Li, Q., Tang, S., Xiao, Y., & Wu, J. (2020). An updated estimating of the risk of transmission of the novel coronavirus (2019-nCov). *Infectious Disease Modelling*. 5: 248-255.

Taylor, R., Kotian, P., & Warrant, T., et al., (2016). BCX4430-a broad-spectrun antiviral adenosine nucleoside analog under development for the treatment of Ebola virus disease. *J Infect Public Health*. 9: 220-226.

Thomas, P. A. (2003). Severe acute respiratory syndrome (SARS): a bolt from the blue. *Indian Journal of Medical Microbiology*. 21(3): 150-151.

Tian, S., Hu, N., Lou, J., Chen, K., Kang, X., Xiang, Z., Chen, H., Wang, D., Liu, N., Liu, D., Chen, G., Zhang, Y., Li, D., Li, J., Lian, H., Niu, S., Zhang, L., & Zhang, J. (2020). Characteristics of COVID-19 infection in Beijing. *Journal of Infection*. 80: 401-406.

Tong, T. R. (2006). Severe acute respiratory syndrome coronavirus (SARS-CoV). *Perspectives in Medical Virology*. 16: 43-95.

Tseng, C. T., Sbrana, E., Iwata-Yoshikawa, N., Newman, P. C., Garron, T., Atmar, R. L., Peters, C. J., & Couch, R. B. (2012). Immunization with SARS coronavirus vaccines leads to pulmonary immunopathology on challenge with SARS virus. *PLoS ONE*. 7(4): e35421.

Tsui, W. M. S. (2003). Coronavirus is the cause of the severe acute respiratory syndrome (SARS) outbreak in Hong Kong and worldwide. *Advances in Anatomic Pathology*. 10(4): 236.

Van Boheemen, S., de Graaf, M., Lauber, C., Bestebroer, T. M., Raj, V. S., Zaki, A. M., Osterhaus, A. D., Haagmans, B. L., Gorbalenya, A. E., Snijder, E. J., & Fouchier, R. A. (2012). Genomic characterization of a newly discovered coronavirus associated with acute respiratory distress syndrome in humans. *MBio 3*.

Volz, A., Kupke, A., Song, F., Jany, S., Fux, R., & Shams-Eldin, H., et al, (2015). Protective efficacy of recombinant modified vaccinia virus Ankara delivering Middle East respiratory syndrome coronavirus spike glycoprotein. *J Virol*. 89(16): 8651-8656.

Wall, A. C., Park, Y. J., Tortoici, M. A., Wall, A., McGuire, A. T., & Veesler, D. (2020). Structure, function, and antigenicity of the SARS-CoV-2 spike glycoprotein. *Cell.* http://doi.org/10.1016/ j.cell.2020.02.058,

Wang, W., & Ruan, S. (2004). Simulating the SARS outbreak in Beijing with limited data. *Journal of Theoretical Biology*. 227: 369-379.

Wang, Z. G., Xu, S. P., & Zhang, Y. J. (2006). The possible origin of recent human SARS coronavirus isolate from China. *Acta Virologica*. 50(3): 211-213.

Wang, H., Rao, S., & Jiang, C. (2007). Molecular pathogenesis of severe acute respiratory syndrome. *Microbes and Infection*. 9: 119-126.

Wang, N., Shi, X., Jiang, L., Zhang, S., Wang, D., & Tong, P., et al., (2013). Structure of MERS-CoV spike receptor-binding domain complexed with human receptor DPP4. *Cell Res*. 23(8): 986-993.

Wang, L., Shi, W., Joyce, M. G., Modjarrad, K., Zhang, Y., & Leung, K, et al., (2015). Evaluation of candidate vaccine approaches for MERS-CoV. *Nat Commun*. 6: 7712.

Wang, Q., Wong, G., Lu, G., Yan, J., & Gao, G. F. (2016). MERS-CoV spike protein: targets for vaccines and therapeutics. *Antiviral Research*. 133: 165-177.

Wang, Y., Sun, J., Zhu, A., Zhao, J., & Zhao, J. (2018). Current understanding of Middle East respiratory syndrome coronavirus infection in human and animal models. *Journal of Thoracic Disease*. 10(19): S2260-S2271.

Wang, C., Horby, P. W., Hayden, F. G., & Gao, G. F. (2020). A novel coronavirus outbreak of global health concern. *Lancet* (2020 Jan 29). Epub ahead of print, http://www.thelancet.com/journals/lancet/ article/PIIS0140-6736(20)30185-9/full text.

Wang, K., Kang, S., Tian, R., Zhang, X., Zhang, X., & Wang, Y. (2020). Imaging manifestations and diagnostic value of chest CT of coronavirus disease 2019 (COVID-19) in the Xiaogan area. *Clinical Radiology*. http://doi.org/10.1016/j.crad.2020.03.004.

Wang, M., Cao, R., Zhang, L., Yang, X., Liu, J., & Xu, M., et al., (2020). Remdesivir and chloroquine effectively inhibit the recently emerged novel coronavirus (2019-nCoV) *in vitro*. *Cell Res*. 1-3.

Wenzel, R. P., Bearman, G., & Edmond, M. B. (2005). Lessons from severe acute respiratory syndrome (SARS): implications for infection control. *Archives of Medical Research*. 36: 610-616.

Wernery, U., Lau, S. K. P., & Woo, P. C. Y. (2017). Middle East respiratory syndrome (MERS) coronavirus and dromedaries. *The Veterinary Journal*. 220: 75-79.

White, J. M., Delos, S. E., Brecher, M., & Schornberg, K. (2008). Structures and mechanisms of viral membrane fusion proteins: multiple variations on a common theme. *Crit. Rev. Biochem. Mol. Biol*. 43: 189-219.

White, J. M., & Whittaker, G. R. (2016). Fusion of enveloped viruses in endosomes. *Traffic*. 17: 593-614.

Widagdo, W., Okba, N. M. A., Raj, V. S., & Haagmans, B. L. (2017). MERS-coronavirus: From discovery to intervention. *One Health*. 3: 11-16.

Woodhead, M., Ewig, S., and Torres, A. (2003). Severe acute respiratory syndrome (SARS). *Eur Respir J*. 21: 739-740.

Wong, C. K. K., Lai, V., & Wong, Y. C. (2012). Comparison of initial high resolution computed tomography features in viral pneumonia between metapneumovirus infection and severe acute respiratory syndrome. *European Journal of Radiology*. 81: 1083-1087.

Wong, H. H., Fung, T. S., Fang, S., Huang, M., Le, M. T., & Liu, D. X. (2018). Accessory proteins 8b and 8ab of severe acute respiratory syndrome coronavirus suppress the interferon signaling pathway by mediating ubiquitin-dependent rapid degradation of interferon regulatory factor 3. *Virology*. 515: 165-175.

World Health Organization. (2015). *Middle East respiratory syndrome coronavirus (MERS-CoV): Thailand. Disease outbreak news*. World Health Organization; 2015. Available from: http://www.who.int/csr/don/20-june-2015-mers-thailand/en/.

World Health Organization. (2016). *Middle East respiratory syndrome coronavirus [MERS-CoV]*; 2016. (http://www,who.int/emergencies/mers-cov/en/).

Wu, G., and Yan, S. (2004). Potential targets for anti-SARS drugs in the structural proteins from SARS related to coronavirus. *Peptides*. 25(6): 901-908.

Wu, A., Peng, Y., Huang, B., Ding, X., Wang, X., Niu, P., Meng, J., Zhu, Z., hang, Z., Wang, J., Sheng, J., Quan, L., Xia, Z., Tan, W., Cheng, G., & Jiang, T. (2020). Genome composition and divergence of the novel coronavirus (2019-nCoV) originating in China. *Cell Host and Microbe*.

Xie, C., Jiang, L., Huang, G., Pu, H., Gong, B., Lin, H., Ma, S., Chen, X., Long, B., Si, G., Yu, H., Jiang, L., Yang, X., Shi, Y., & Yang, Z. (2020). Comparison of different samples for 2019 novel coronavirus detection by nucleic acid amplification tests. *International Journal of Infectious Diseases*. 93: 264-267.

Xu, C., Wang, J., Wang, L., & Cao, C. (2014). Spatial pattern of severe acute respiratory syndrome in-out flow in 2003 in mainland China. *BMC Infectious Diseases*. doi: 10.1186/s12879-014-0721-y.

Yam, L. Y. C., Lau, A. C. W., Lai, F. Y. L., Shung, E., Chan, J., & Wong, V. (2007). Corticosteroid treatment of severe acute respiratory syndrome in Hong Kong. *Journal of Infection*. 54: 28-39.

Yang, C. W., & Chen, M. F. (2020). Composition of human-specific slow codons and slow di-codons in SARS-CoV and 2019-nCoV are lower than other coronaviruses suggesting a faster protein synthesis rate of SARS-CoV and 2019-nCoV. *Journal of Microbiology, Immunology and Infection*. http://doi.org/10.1016/j.jmii.2020.03.002.

Yavarian, J., Shafiei Jandaghi, N. Z., Naseri, M., Hemmati, P., Dadras, M., Gouya, M. M., & Azad, T. M. (2018). Influenza virus but not MERS coronavirus circulation in Iran, 2013-2016: comparison between pilgrims and general population. *Travel Medicine and Infectious Disease*. 21: 51-55.

Yoon, J. H., Lee, J. Y., Lee, J., Shin, Y. S., Jeon, S., Kim, D. E., Min, J. S., Song, J. H., Kim, S., Kwon, S., Jin, Y. H., Jang, M. S., Kim, H. R., &

Park, C. M. (2019). Synthesis and biological evaluation of 3-acyl-2-phenylamino-1,4-dihydroquinolin-4(1H)-one derivatives as potential MERS-CoV inhibitors. *Bioorganic and Medicinal Chemistry Letters*. 29: 126727.

Yount, B., Curtis, K. M., Fritz, E. A., Hensley, L. E., Jahrling, P. B., Prentice, E., Denison, M. R., Geisbert, T. W., & Baric, R. S. (2003). Reverse genetics with a full-length infectious cDNA of severe acute respiratory syndrome coronavirus. *PNAS*. 100(22): 12995-13000.

Yu, P., Hu, B., Shi, Z. L., & Cui, J. (2019). Geographical structure of bat SARS-related coronaviruses. *Infection, Genetics and Evolution*. 69: 224-229.

Yu, F., Du, L., Ojcius, D. M., Pan, C., & Jiang, S. (2020). Measures for diagnosing and treating infections by a novel coronavirus responsible for a pneumonia outbreak originating in Wuhan, China. *Microbes and Infection*. 22: 74-79.

Yu Jun, I. S., Anderson, D. E., Zheng Kang, A. E., Wang, L. F., Rao, P., Young, B. E., Lye, D. C., & Agrawal, R. (2020). Assessing viral shedding and infectivity of tears in coronavirus disease 2019 (COVID-19) patients. *Ophthalmology*. Doi: http://doi.org/10.1016/j.ophtha.2020.03.026.

Yuan, Y., Cao, D., Zhang, Y., Ma, J., Qi, J., & Wang, Q., et al., (2017). Cryo-EM structures of MERS-CoV and SARS-CoV spike glycoproteins reveal the dynamic receptor binding domains. *Nat Commun*. 8: 15092.

Yuan, M., Yin, W., Tao, Z., Tan, W., & Hu, Y. (2020). Association of radiologic findings with mortality of patients infected with 2019 novel coronavirus in Wuhan. *PLOS ONE*. 15(3): e0230548. http://doi.org/10.1371/journal.pone.0230548.

Zaki, A. M., van Boheemen, S., Bestebroer, T. M., Osterhaus, A. D., & Fouchier, R. A. (2012). Isolation of a novel coronavirus from a man with pneumonia in Saudi Arabia. *N Engl J Med*. 367: 1814-1820.

Zhai, S., Liu, W., & Yan, B. (2007). Recent patents on treatment of severe acute respiratory syndrome (SARS). *Recent Patents on Anti-Infective Drug Discovery*. 2(1): 1-10.

Zhang, N., Jiang, S., & Du, L. (2014). Current advancements and potential strategies in the development of MERS-CoV vaccines. *Expert Rev. Vaccines*. 13: 761-774.

Zhang, N., Tang, J., Lu, L., Jiang, S., & Du, L. (2015). Receptor-binding domain-based subunit vaccines against MERS-CoV. *Virus Res*. 202: 151-159.

Zhang, S., Zhou, P., Wang, P., Li, Y., Jiang, L., Jia, W., Wang, H., Fan, A., Wang, D., Shi, X., Fang, X., Hammel, M., Wang, S., Wang, X., & Zhang, L. (2018). Structural definition of a unique neutralization epitope on the receptor-binding domain of MERS-CoV spike glycoprotein. *Cell Reports*. 24: 441-452.

Zhang, Y. Z., & Holmes, E. C. (2020). A genomic perspective on the origin and emergence of SARS-CoV-2. *Cell*. http://doi.org/10.1016/j.cell.2020.03.035

Zhang, D. H., Wu, K. I., Zhang, X., Deng, S. Q., & Peng, B. (2020). *In silico* screening of Chinese herbal medicines with the potential to directly inhibit 2019 novel coronavirus. *Journal of Integrative Medicine*. 18: 152-158.

Zhao, S., Ling, K., Yan, H., Zhong, L., Peng, X., Yao, S., Huang, J., and Chen, X. (2020). Anesthetic management of patients with COVID 19, infections during emergency procedures. *Journal of Cardiothoracic and vascular anesthesia*. 34: 1125-1131.

Zheng, B., He, M. L., Wong, K. L., Lum, C. T., Poon, L. L. M. Peng, Y., Guan, Y., Lin, M. C. M., & Kung, H. F. (2004). Potent inhibition of SARS-associated coronavirus (SCOV) infection and replication by type I interferons (IFN-alpha/beta) but not by type II interferon (IFN-gamma). *J. Interferon Cytokine Res. Off. J. Int. Soc. Interferon Cytokine Res*. 24: 388-390.

Zheng, B., Cao, K. Y., Chan, C. P. Y., Choi, J. W. Y., Leung, W., Leung, M., Duan, Z. H., Gao, Y., Wang, M., Di, B., Hollidt, J. M., Bergmann, A., Lehmann, M., Renneberg, I., Tam, J. S. L., Chan, P. K. S., Cautherley, G. W. H., Fuchs, D., & Renneberg, R. (2005). Serum neopterin for early assessment of severity of severe acute respiratory syndrome. *Clinical Immunology*. 116: 18-26.

Zhong, N. S. (2003). The clinical diagnosis and treatment of SARS at present. *Zhong Guoyi Xuelun Tanbao*. 4: 29.

Zhong, N. S., & Wong, G. W. K. (2004). Epidemiology of severe acute respiratory syndrome (SARS): adults and children. *Paediatric Respiratory Reviews*. 5: 270-274.

Zhou, Y., Yang, Y., Huang, J., Jiang, S., & Du, L. (2019). Advances in MERS-CoV vaccines and therapeutics based on the receptor-binding domain. *Viruses*. 11: 60.

Zumla, A., Chan, J. F. W., Azhar, E. I., Hui, D. S. C., and Yuen, K. Y. (2016). Coronaviruses-drug discovery and therapeutic options. *Nat Rev Drug Discov*. 15: 327-347.

Zhang, X. M., Kousoulas, K. G., & Storz, J. (1992). The hemagglutinin/esterase gene of human coronavirus strain OC43: phylogenetic relationships to bovine and murine coronaviruses and influenza C virus. *Virology*. 186: 318-323.

Zhang, R., Wang, K., Lv, W., Yu, W., Xie, S., Xu, K., Schwarz, W., Xiong, S., & Sun, B. (2014). The ORF4a protein of human coronavirus 229E functions as a viroporin that regulates viral production. *Biochimica et Biophysica Acta*. 1838: 1088-1095.

Zhang, Y., Li, J., Xiao, Y., Zhang, J., Wang, Y., Chen, L., Paranhos-Baccala, G., Ren, L., & Wang, J. (2015). Genotype shift in human coronavirus OC43 and emergence of a novel genotype by natural recombination. *Journal of Infection*. 70: 641-650.

Zhao, Q., Li, S., Xue, F., Zou, Y., Chen, C., Bartlam, M., & Rao, Z. (2008). Structure of the main protease from a global infectious human coronavirus, HCoV-HKU2. *Journal of Virology*. 82(17): 8647-8655.

About the Authors

Mohamad Hesam Shahrajabian
Biotechnology Research Institute,
Chinese Academy of Agricultural Sciences,
Beijing, China

He is a senior researcher of Agronomy and Biotechnology. He is interested in crops and herb which are related to traditional Medicine, especially Chinese and Iranian traditional Medicine crops relating to organic farming and sustainable agriculture. His current research is the influence of medicinal herbs and fruits on human coronaviruses. His full profile is available in http://orcide.org/0000-0002-8638-1312.

Corresponding E-mail: hesamshahrajabian@gmail.com.

Wenli Sun
Biotechnology Research Institute,
Chinese Academy of Agricultural Sciences,
Beijing, China

She is an associate professor working on related topic of traditional Chinese medicine, allelopathic influence and sustainable agriculture. She is also working on topics which are related to Biotechnology and Molecular Science. Her current research is survey on history of human coronavirus and influene of traditional Chinese medicine in prevention and treatment of human coronavirus. Her full profile is available in http://orcide.org/0000-0002-1705-2996.

Corresponding E-mail: sunwenli@caas.cn.

Qi Cheng

Biotechnology Research Institute, Chinese Academy of Agricultural Sciences, Beijing, China

College of Life Sciences, Hebei Agricultural University, Baoding, Hebei, 071000, China; Global Alliance of HeBAU-CLS&HeQiS for BioAl-Manufacturing, Baoding, Hebei 071000, China

He is a professor of Biotechnology and his researches have connected with agrobiotechnology and agrotechnology. Presently, he is interested to traditional Chinese medicine and molecular researches; also he is interest in doing researches on human coronaviruses and the impact natural products for treatment and prevention of SARS, MERS and SARS-CoV-2. His full profile is available in http://orcide.org.0000-0003-1269-6386.

Corresponding E-mail: chengqi@caas.cn.

INDEX

T

V

W